SUPERCOMPUTERS

This is one of a series of books on technology and entrepreneurship sponsored by the IC^2 Institute at The University of Texas at Austin, under the general editorship of Raymond W. Smilor and Robert L. Kuhn. The series provides materials from practical and scholarly perspectives that can in turn assist leaders in policy and decision making. It ties practice to theory. Each book in the series is designed to build on the one before it so that an extensive foundation of research and practical insights are available to leaders in business, government and academia. In the process, the series provides foresight into complex topics. It offers an opportunity to introduce relevant topics in technology and entrepreneurship as well as allow for important contributions to established lines of research and practice.

OTHER TITLES IN THE SERIES

Commercializing Defense-Related Technology, edited by Robert L. Kuhn

Corporate Creativity: Robust Companies and the Entrepreneurial Spirit, edited by Raymond W. Smilor and Robert L. Kuhn

Technology Venturing: American Innovation and Risk Taking, edited by Eugene B. Konecci and Robert L. Kuhn

Managing Take-off in Fast Growth Companies, edited by Raymond W. Smilor and Robert L. Kuhn

Commercializing SDI Technologies, edited by Stewart Nozette and Robert Lawrence Kuhn

SUPERCOMPUTERS

A Key to U.S. Scientific, Technological, and Industrial Preeminence

**EDITED BY J. R. KIRKLAND
AND J. H. POORE**

 PRAEGER

New York
Westport, Connecticut
London

Library of Congress Cataloging-in-Publication Data

Supercomputers : a key to U.S. scientific,
  technological, and industrial preeminence.

  Bibliography: p.
  Includes index.
  1. Supercomputers.   2. Supercomputers—United
States.   I. Kirkland, J. R.   II. Poore, J. H.
QA76.5.S89474   1987   004.1'1   87-11808
ISBN 0-275-92622-2 (alk. paper)

Library of Congress Catalog Card Number: 87-11808
ISBN: 0-275-92622-2

First published in 1987

Praeger Publishers, One Madison Avenue, New York, NY 10010
A division of Greenwood Press, Inc.

Printed in the United States of America

∞

The paper used in this book complies with the Permanent
Paper Standard issued by the National Information Standards
Organization (Z39.48-1984).

10   9   8   7   6   5   4   3   2   1

CONTENTS

Part II: Perspectives of Supercomputer Manufacturers

Part III: The Role and Scope of Federal Agencies in Supercomputer Development

Part IV: Supercomputers and the University

Part V: Industrial Applications of Supercomputer Technology

LIST OF FIGURES

LIST OF TABLES

PREFACE: SUPERCOMPUTERS, SCIENCE, AND HUMAN PURPOSE

In his keynote speech to the conference, Supercomputers: A Key to U.S. Scientific, Technological, and Industrial Preeminence, held in Tallahassee, Florida, June 9–11, 1985, Senator Albert Gore of Tennessee recited the names of Darwin, Galileo, Einstein, and Freud as examples of scientists whose major discoveries redefined the relationship of people to the world they inhabit. Gore also prophesied that new generations of supercomputers will be the instruments used to attain even greater heights of scientific achievement, and to assure humankind's global future. There is no doubt that Senator Gore's vision of our future is correct. We therefore need only decide what we need to do to get there. This volume seeks not only to provide insights on the current state of supercomputers, but also to suggest important directions for their development applications and commercialization.

Certainly, we will need supercomputers, but that is not all. We will also need, as soon as possible, a few more Darwins, Galileos, Einsteins, and Freuds. We must also find a few more Francis Bacons—a few people like that great scientist, philosopher, and statesman of Elizabethan England.

Unlike Darwin, Galileo, Einstein, and Freud, Sir Francis Bacon found his real genius not as a practitioner of science but as a statesman of science. And genius it was, for his statesmanship changed the nature of scientific endeavor by emphasizing that science is a social activity, properly conducted by a great community of scholars working together to achieve common human goals. He was the man who first saw science as a methodology, a spirit of inquiry, a logical tool that would serve as "an invention of inventions." He was the man whose thinking laid the intellectual groundwork for the national and international research programs that are common in the twentieth century. Bacon believed that nature could be mastered only by means of a comprehensive research program implemented on a very large scale. He had a seer's vision of what science could accomplish by coordinating the activities of many researchers working in a single scientific tradition, and by initiating government-sponsored research for those projects that were beyond private means and that could not be accomplished "in the hourglass of one man's life."

What were these things that needed to be done? The Baconian program was not "science for its own sake"; it was a program aimed at attaining knowledge to be used for power over nature for the improvement of the human lot. For Bacon, science was strictly "for the uses of life," and he cautioned against knowledge without "charity"—knowledge without human purpose.

What would some new Francis Bacons be saying today? What would be the essential topics of conversation for these new great statesmen of science? They would be exactly the same as they were for Francis Bacon, and their theme would be exactly the same as his: the advancement of knowledge for the attainment of power over nature, along with the employment of that power for the improvement of human life. The main difference would be the use of a term Bacon himself had not heard: "supercomputer." Otherwise, the agenda would be the same: the solution of the world's problems through a coherent, comprehensive, coordinated, and very large-scale enterprise of the entire scientific community using the most powerful intellectual tool available, a tool to facilitate the "invention of inventions" to achieve a better world.

The June, 1985 conference at Florida State University brought together scientists from government, industry, and the academic world to examine the role of these new tools called supercomputers. The specific objectives of the conference were to

- assess current application of supercomputer technology to industry and future needs, and to suggest a research agenda for further development;
- review the education and training needs for supercomputer technology in industry and the academic community;
- suggest coalitions of academic, economic, and governmental resources to push forward a research and training agenda for supercomputer technology;
- provide recommendations for federal policies to enhance supercomputer technology and applications through a hearing scheduled by the U.S. House of Representatives Committee on Science and Technology.

The chapters in this volume are an outgrowth of that conference. They provide a perspective on the current status and future direction of supercomputer technology and applications from the viewpoints of policy makers, scientists, industrial users, and manufacturers. Their recommendations for action are synthesized in a series of initiatives that should be considered taken by government, the private sector, and universities to secure the preeminence of the United States in supercomputer technology.

ACKNOWLEDGMENTS

Numerous persons and organizations contributed to this volume. The editors would like to acknowledge and thank them all.

First, thanks should go to the sponsors of the conference, Supercomputers: A Key to U.S. Scientific, Technological, and Economic Preeminence, held at Florida State University, June 9–11, 1985. Generous contributions from the RGK Foundation, Peat Marwick Mitchell, ETA Systems, Cray Research, Florida State University, Westinghouse Corporation, and the IC^2 Institute at The University of Texas at Austin made possible the symposium that produced the recommendations and chapters in this volume.

Key to the conception and development of the conference were Representative Don Fuqua, Chairman of the U.S. House of Representatives Committee on Science and Technology, and members of his senior staff, Dr. Harold Hanson and Dr. Grace Ostenso.

Dr. Robert Johnson and Dr. Joseph Lannutti of Florida State University contributed freely of their time to both the development of the conference agenda and the chapters in this volume. Dr. James Decker, Deputy Director of the Office of Energy Research, Department of Energy, and Dr. John Connelly, Director of Advanced Scientific Computing for the National Science Foundation, helped identify key persons in the scientific community and the government agencies working on the leading edge of research in large-scale computational systems. Ronya Kozmetsky, president, and Cynthia Smith, administrator, of the RGK Foundation provided invaluable logistical support to the conference.

We would like to express special thanks to Dr. George Kozmetsky and Dr. Raymond Smilor of the IC^2 Institute at The University of Texas for their advice and assistance with the preparation of this volume. As always, their suggestions have brought greater clarity to a complex subject.

We wish to thank Patsy King of the Georgia Institute of Technology, Myrna Braziel of the RGK Foundation, and Linda Teague, Nancy Richey, and Becky Younger of the IC^2 Institute for their assistance with the preparation of this volume for publication. Their skills, patience, and occasional coffee cakes sustained us during the project.

INTRODUCTION

Supercomputers are the fastest and most powerful scientific computing systems available to solve large-scale scientific problems; they offer speed and capacity far greater than machines available primarily for commercial use. Supercomputers, including the hardware, software, supporting peripherals, and the facilities and personnel needed for their appropriate use, have been a crucial part of the advances in science, engineering, national security, and industrial productivity over the past 20 years. Until the beginning of the 1980s, the United States was clearly the world leader in advanced computer technology.

The preeminence of the United States in large-scale computing has been a result of the confluence of three factors: the vitality of the U.S. computer industry, the far-sighted policies of the federal government, and the leadership of scientists and engineers from universities and government laboratories. The Atomic Energy Commission, at the urging of John von Neumann, initiated the use of large-scale computation in research and weapons design; NASA advanced the use of supercomputing in its scientific programs. U.S. universities and government laboratories conducted the research that formed the basis for constructing and applying computers, trained the needed scientific and engineering personnel, and made computers and computing an essential tool in scientific and engineering research.

The federal government vigorously implemented policies that supported these efforts, granted generous funds for computation, and—through its role as the major purchaser of scientific computers—provided the incentives and insured the market for these unique machines. Forward-looking corporations exploited the scientific and engineering opportunities, developed an advanced industrial technology, and created this most vital component of the U.S. economy.

During the 1970s the federal government retreated from its support of large-scale computing in universities. The National Science Foundation (NSF) program to provide the expanded university computing facilities for scientific and engineering research was terminated in 1972; at about the same time IBM discontinued its generous discounts for the purchase of computing equipment by academic institutions as a result of pressures from the Justice Department and competitors. Large-scale university computing facilities withered while the action shifted to national laboratories and to industrial users.

As U.S. investment in advanced computational technology declined, foreign technology—especially Japanese—made a quantum leap forward. Faced with the prospect of becoming a second-tier nation in world technology, with the implications for military and economic security such a position would hold, Congress and the administration launched an initiative to reestablish our dominant position in computer technology.

The most well-known initiative taken by the federal government was the appointment of the Lax Committee, which was charged with developing recommendations to secure U.S. technological dominance of the large-scale computer field.

The Lax Panel, sponsored by the Department of Defense (DOD) and NSF in cooperation with NASA and the Department of Energy (DOE), identified two major problems: "important segments of the research and defense communities lack effective access to supercomputers; and current plans for supercomputer development can only achieve a fraction of the capability and capacity technically achievable in this decade." The panel recommended the establishment of a national program to stimulate exploratory development and expanded use of advanced computer technology, with four components:

1. Increased access for the scientific and engineering research community through high bandwidth networks to adequately and regularly updated supercomputing facilities and experimental computers;

2. Increased research in computational mathematics, software, and algorithms necessary to the effective and efficient use of supercomputer systems;

3. Training of personnel in scientific and engineering computing; and

4. Research and development basic to the design and implementation of new supercomputer systems of substantially increased capability and capacity, beyond that likely to arise from commercial requirements alone.

As a result of the Lax Committee's recommendations, Congress has actively supported programs for development of supercomputer technology, and agencies concerned with such development have formed the Federal Coordinating Council on Science, Engineering, and Technology, to foster exchange of technology and resources.

The outgrowth of the increased support at the national level has been the establishment of new supercomputer centers at selected universities. These centers, for the most part, have been founded and staffed through joint ventures that combine resources of federal and state governments, universities, and the business community. In the past two years such centers have been established at Florida State University, in conjunction with the DOE, and NSF has funded centers at Cornell University, the University of California at San Diego, the University of Illinois, and a center for a consortium of northeastern universities to be located at Princeton University. These centers join pioneering efforts at the University of Minnesota, the University of Georgia, Purdue University, and Colorado State University.

Concurrent with the new federal initiatives to establish supercomputer centers for academic use and basic scientific research, the past five years have seen a rapidly growing use of supercomputers by industry. Oil companies have turned to supercomputers for assistance with exploration for new deposits of oil and gas reserves and for tertiary recovery of oil deposits. Manufacturers of advanced microelectronic circuits, involving hundreds of thousands of logical elements on a chip, are using supercomputers to check circuit designs for completeness and consistency. Airplane manufacturers have used supercomputers for many years to assist with the analysis of aerodynamic flow around proposed designs and have found that computer analysis can greatly reduce the very expensive wind-tunnel testing time required to perfect new designs. Even moviemakers and television have turned to supercomputers to create the elaborate visual effects that have appeared on recent films. Since the private sector has offered a new and substantial market for supercomputers in the past few years, it is expected that applications will grow in number as the computing power of supercomputers increases to permit more realistic simulation of complex, three-dimensional, time-dependent phenomena.

Even with the expanded use by industry and the establishment of university centers, inadequate access to supercomputers for all disciplines is a major concern of researchers who require large-scale computing capabilities to sustain their work.

One of the most significant problems is cost. Supercomputer procurement costs run from $10 to $20 million with comparable operating costs for each year of operation. This level of investment can be made by larger members of industry, some areas of the federal government, and a few academic institutions. However, the high price tag is not the only factor that limits access. Operating a supercomputer requires highly trained individuals, well-supported software for user operation and research, and time for system experimentation.

These required investments in time and money often combine to limit access for the university research community and small businesses. They can also affect industrial and other access to future generations of supercomputers by making the payback period for the present generation of machines longer than its technological life.

Linking or networking and sharing resources are the best means of reducing costs. Three possible approaches for reliable and efficient supercomputer access have been suggested; all require a nationwide interdisciplinary network:

1. to enhance supercomputer capacity and staff at existing centers, such as those DOE and NSF support;

2. to provide supercomputers to selected government laboratories without such facilities and make them available to the broad research and development (R&D) community through networking; and

3. to establish additional regional centers at selected universities, interconnected with existing facilities at other universities and government laboratories.

In addition to the need to expand to supercomputers, there is a major need to push forward the frontiers of very large-scale integrated circuits (VLSIC) technology. Even with the increased usage and expanded applications, the development of supercomputers, as now planned in the United States, will yield only a small fraction of the capability and capacity thought to be technically achievable in this decade.

Significant new research and development effort is necessary to overcome technological barriers to the creation of a generation of supercomputers that tests these technical limits. Computer manufacturers in the United States have neither the financial resources nor the commercial motivation in the present market to undertake the requisite exploratory research and development without partnership with government and universities. These partnerships of business-government and academia are termed "technology venturing," as defined by George Kozmetsky:

Technology venturing is an entrepreneurial process by which major institutions and individuals take and share risk in integrating and commercializing scientific research and various technologies. It incorporates a dynamic private sector, an effective capital venture industry, and creative role for government and new academic relationships. Technology venturing is essential in improving an educational structure, fulfilling critical manpower requirements, and enhancing our industrial creativity and innovation. It is the primary means for encouraging the emergence of a myriad of technology related businesses . . . in the context of a private enterprise system that has always been the unique American way to achieve and maintain U.S. economic, scientific, and technological preeminence.[1]

The need to marshal and link resources was the principal theme that emerged from the Florida State conference. Speakers and participants were unanimous in calling for new lines of communication between technology and computer centers and for new, imaginative relationships between the business community and scientists in federal laboratories. The recommendations and discussions in this volume reflect ideas on means to affect these new relationships.

J. R. Kirkland
J. H. Poore

NOTE

1. For an in-depth discussion of this concept, see Eugene Konecci and Robert Kuhn, eds., *Technology Venturing* (New York: Praeger, 1985).

PART I

NATIONAL ISSUES AND PERSPECTIVES

1

INITIATIVES FOR POLICIES AND PROGRAMS

J. R. KIRKLAND and J. H. POORE

Several specific initiatives to enhance supercomputer technology and applications emerged from papers and discussions at the Florida State University conference. These recommendations centered around four major themes:

1. The need to establish a national, standardized network to link supercomputers to take maximum advantage of scientific resources in government, universities, and industry

2. The need for a consistent, national policy that would reduce administrative and regulatory barriers that limit access and inhibit the exchange of scientific information among government-funded supercomputer centers, university-based scientists, and industrial users of supercomputers

3. The need for national guidelines to promote development of compatible or standard software and peripheral systems for the rapidly expanding super-computer "community"

4. The need for a national initiative taken by federal and state governments, industry, and universities to develop faculty and students trained in the use and applications of supercomputer technology

NETWORKING

The strongest recurring theme throughout the Florida State conference was the call for a high-function data-image communications network between government-funded supercomputer centers and those in industry. Strong arguments were made that a single such network should serve the university community, the national laboratories, and industrial laboratories. Further examination of the situation indicates that the need is not limited to data communications between supercomputers and remote users, but extends to the ubiquitous data communications among computers of all sizes and purposes, to general-image transmission, and to interactive communications.

Unless strong leadership is exerted soon, the United States will have a multitude of independently designed networks by the Department of Energy, the National Science Foundation, NASA, the National Bureau of Standards (NBS), and others. These networks will not be connected effectively, will not have compatible characteristics, will not be connected to the university campuses in a system to promote facile exchange of scientific data, and will not foster exchange with industrial or federal laboratories.

Initiative 1: The House Committee on Science and Technology should be asked to hold hearings on the subject of a national network servicing the entire science community for voice-data-image networking and the committee should identify and designate a lead agency to design, build, and operate a national network to link the various elements of the emerging supercomputer community.

Initiative 2: A committee of university and private-sector users of supercomputers should be formed to advise on the planning, design, and operation of a national scientific network. While "unofficial," the committee should offer the congressional committees and federal agencies direct and specific recommendations for improving the exchange of information between scientists at various supercomputer centers. The committee should be funded from private sources and not be dependent on government for support.

Initiative 3: A conference should be held on the subject of creating a national supercomputer network. This conference should be organized by the committee recommended in Initiative 2 and sponsorship should be sought from the private sector with full participation by congressional committees and federal agencies.

Initiative 4: Individual universities, regional consortia, and professional societies should prepare specific case studies outlining their needs for a national network for supercomputer users. These cases should be oriented toward the potential impact of ubiquitous communications on the development of universities and the various scientific disciplines. Studies should be presented to the private-sector committee and to relevant congressional and federal agency committees.

Initiative 5: The Office of Science and Technology Policy should call upon all federal agencies to begin revising programs and plans to make existing and planned networks and positions more compatible and to begin removing incompatibilities from these networks.

REMOVAL OF ADMINISTRATIVE IMPEDIMENTS TO EFFECTIVE USE OF SUPERCOMPUTER FACILITIES

A number of circumstances exist that either cause available supercomputers to go unused for a significant amount of time, or that impose unnecessarily awkward logistical problems for scientists. These barriers could be broken down through appropriate administrative action.

For example, a scientist at the University of Minnesota performing NSF-sponsored research in atmospheric science might be required by agency regulations to use a supercomputer in Colorado, even though a similar computer is available at the home campus or at a site to which the home campus has a better data-communications link. A colleague, also at Minnesota, with similar research sponsored by NASA, may be required to use a different computer at yet another laboratory. The research of these two scientists at the same university could be greatly impeded by bureaucratic regulations simply because their work is sponsored by different agencies of government.

This situation is even more burdensome for universities that own their supercomputers, because government agencies are reluctant to reimburse the computer costs at facilities other than their own agency-sponsored laboratories.

Yet another example exists where scientists are restricted by regulation from using resources in private industry. If a supercomputer is owned by a corporation and there is excess time available on the computer, there is no mechanism for the excess to be used by university researchers who need the supercomputer time, even if the computer is fully funded by federal contracts. For example, if Westinghouse wishes to offer computer time to Carnegie-Mellon or the University of Pittsburgh, there is no ready administrative mechanism to facilitate such use of resources.

Initiative 6: Fiscal and administrative directives should be initiated by the Office of Management and Budget to facilitate the feeding of idle capacity of unclassified, federally funded supercomputers into a grid of availability. University scientists should have access to this grid for work on federally funded grants, without regard to the funding agency.

Initiative 7: Federal agencies that sponsor supercomputer centers should establish a clearinghouse for usage, including university-owned facilities, so that the most convenient and reasonable logistical arrangements are available to scientists. Universities should establish programs that assist scientists to gain access to the proposed grid of supercomputer centers.

Initiative 8: The private-sector committee (suggested in Initiative 2 to oversee networking) should identify instances where the removal of administrative barriers would result in more supercomputer facilities being made available for research. Universities with proximity to industrial supercomputer sites or with existing collaboration with industry should document such instances in order to facilitate the administrative changes.

Initiative 9: The Department of the Treasury and the appropriate congressional committees should consider tax and other incentives for private industry to make excess capacity of research facilities available to university and government researchers. Such measures should result in expediting access to existing super-computer capacity, promote further exchange between academia and industry, and reduce the need for federal outlays for additional supercomputer centers.

COMPATIBILITY AND STANDARDIZED FORMAT FOR SOFTWARE AND PERIPHERAL SYSTEMS

The need for standardized, advanced software and compatible systems was a recurring theme throughout the discussions at the super-computer conference. The issue is complex. Pointing to the enormous investment in codes written in FORTRAN language, speakers from industry and government recommended that universities train students in the practical tools for industrial application of supercomputer technology. University scientists replied that industry should have abandoned FORTRAN long ago and that FORTRAN is not a viable vehicle for the enriched environments they seek, nor is it a reasonable vehicle for the instruction and creativity needed by academic programs.

With the advent of new sites for supercomputer activity, and with the surge in research programs that use these machines, there is a clear need to define application environments now to facilitate future exchange of codes, techniques, and data bases. Writers and speakers agree there is a need for standardization in large-applications codes, such as NAS-TRAN, which are widely disseminated and actively used. These persons called for open discussion of the issues of standardization, languages, and operating systems because the demands of the user community are expected to increase.

Initiative 10: The National Bureau of Standards, which already has a standards mission in software, should undertake a "guidelines" mission since it is the most readily available broker between industry and academia. NBS should issue guidelines that would ease the process of carrying the enormous investment in existing FORTRAN application codes forward into more modern environments.

Initiative 11: Universities with industry co-op programs should specifically target support of co-op students from computer-science programs into industry supercomputer labs, and, conversely, companies that use co-op students should create billets for them in their supercomputer labs.

Initiative 12: Professional societies should define applications environments for their fields of science and urge their adoption by the sponsoring agencies so that suppliers failing to meet minimal standards would be discontinued. The environment descriptions might include the following:

- identify the operating systems of the supercomputer that are to be recognized;
- identify software libraries that are to be available;

- state the minimum number of files and file sizes to be made available to users on the supercomputer;
- identify acceptable front-end systems, their operating systems, and the minimum user disk space acceptable;
- identify data-communications means expected at each site;
- suggest tape formats or other means of transferring files from one recognized system to another.

These measures should enhance the ability of scientists to share programs and data and to collaborate on research programs that might have different sponsors.

DEVELOPING HUMAN RESOURCES FOR SUPERCOMPUTER TECHNOLOGY

The lack of personnel trained on supercomputers is a key limitation to full utilization of supercomputers in both universities and industry. While significant advances are continuing in the development of hardware and software, the education and development of scientists capable of utilizing supercomputers to their fullest extent is lagging behind. There is a dearth of both faculty and industrial scientists who understand supercomputer potential, although a number of partnership arrangements exist for the exchange of scientists and students between universities and industry.

Initiative 13: A national clearinghouse program should be established to coordinate government training grants, university sabbatical opportunities, and industrial training opportunities in the interest areas of supercomputer applications, systems software development, and networking. Such a program should provide means of identifying needs and resource availability and serve as a public-private funding mechanism to promote exchange.

Initiative 14: Universities should review their faculty development programs to assure that their faculties have had the opportunity to learn effective use of modern computer facilities and the extensive, available software tools. Many of the new systems—e.g., the computer-aided engineering and design systems—take enormous amounts of time to master and are often beyond self-learning. A similar recommendation would be appropriate for the professional staffs at the national laboratories.

Initiative 15: Regional accrediting boards should maintain high standards in their review of university faculty preparedness to use and to train students in the use of modern computational methods.

Programmatic accrediting boards in certain disciplines have expressed concern in the past about the scarcity of facilities. As facilities become more readily available, the boards should require faculty training for those offering courses on supervising research utilizing supercomputers. Industry should support such requirements since it is the beneficiary of well-trained scientists in the supercomputer field.

2

SUPERCOMPUTERS AND NATIONAL POLICY: MAINTAINING U.S. PREEMINENCE IN AN EMERGING INDUSTRY

GEORGE KOZMETSKY

The supercomputer industry is an emerging industry that has been evolving over the last 25 years. Yet it is still in its infancy. Until 1980 it was predominantly a U.S. development. Since 1980 it has become a fiercely competitive international race for scientific and economic preeminence.

A number of key policy-related issues must be addressed to ensure a robust U.S. supercomputer industry and to enhance its impacts on other industries and society. Among these issues are:

1. How will supercomputer R&D continue to be performed and funded in the future? What are the changing patterns?
2. What kinds of complex applications and needs will drive the advanced developments of supercomputers?
3. Can cooperative research reduce the time required for supercomputer development and maintain the U.S. competitive position in the global marketplace?
4. How can the results of cooperative research in supercomputer technology be transferred and diffused regionally and institutionally?
5. How can supercomputer advances be transferred and diffused to small and medium-sized companies?

BACKGROUND FOR UNDERSTANDING THE ISSUES

The nature of supercomputers is complex, confusing, and rapidly evolving. An "Annotated Bibliography of Literature on Supercomputers," compiled by Dr. James Browne and John Feo of the Department of Computer Sciences and Patricia Roe of the IC^2 Institute, all at The University of Texas at Austin, shows that most of the literature is in terms of applications and architecture. (An unannotated version of this bibliography is included in this book.) The few articles dealing with business and marketing aspects of supercomputers are either company-distributed literature or generalized interviews with selected individuals. There is yet to be developed a body of literature that structures both the academic and professional fields of supercomputers, focuses on policy and social implications, and examines its business and industrial applications. This is another means of confirming that the supercomputer industry is in its infancy.

For policy purposes, it is important to bring some structure to the area of supercomputers along with relevant data and information. The following structure is relevant in understanding the emergence of the supercomputer industry.

• Generations of U.S. supercomputers
• Current structure of the emerging U.S. supercomputer industry
• Current applications, markets, and trends

Generations of U.S. Supercomputers

There is a tendency by all technical persons to discuss supercomputers by generic names or numbers, by performance characteristics, by component contents, or by company names. When an industry is emerging, this adds to the confusion of the public's understanding of the industry's developments. To help clarify the picture, we have structured the generations of supercomputers based on the works of Kai Hwang,[1] Sidney Fernback,[2] and Lloyd Thorndyke,[3] assimilating their data to fit into the categories of generations shown in this table:

Generation	Time Period	Peak Speed
Beginning	1959-60	1-megaflops
First	1970-74	10-100 megaflops
Second	1975-84	100-1000 megaflops
Third	1985-87	2-10 gigaflops
Fourth	1991-	1-Trillion flops or more

Beginnings

Supercomputers started in the late 1950s as R&D projects for scientific-engineering computations in the 1-megaflop range for the Livermore and Los Alamos National Laboratories. IBM and UNIVAC were the prime contractors to deliver what were called the STRETCH and LARC computers. These first machines were "shared risk" developments with the government ensuring the purchase of more than one machine from each vendor. These programs underwent the same problems that so many other early projects experienced. They underestimated the technological complexity of the project. As a consequence their deliveries were late; they incurred substantial financial overruns; and software was minimal. The early commercial supercomputer markets were captured by Control Data Corporation (CDC) with its CDC 1604. IBM and UNIVAC incorporated their gained technology in mainframes for the business and industrial fields.

First Generation

In the early 1970s, government labs were instrumental in starting the first generation of supercomputers to perform as multiprocessors and vector processors. Three manufacturers responded to their requirements: Burroughs, Texas Instruments, and CDC. Burroughs's ILLIAC IV was its only experimental model. It was built at a reported cost of $31 million, and NASA's Ames Research Center was able to achieve the performance levels for which this machine was designed. After more research, Burroughs decided to drop out of the supercomputer race.

Texas Instruments (TI) decided to venture into the supercomputer area by building its advanced scientific computer (ASC). TI built seven systems and delivered six. One went to the Geophysical Fluid Dynamics Laboratory in Princeton, N.J.; one went to the Naval Research Laboratory in Washington, D.C.; a third went to the Ballistic Missile Defense Laboratory in Huntsville, Alabama; the others were for internal use. TI is apparently no longer in the supercomputer field.

CDC developed STAR 100. Four of these were built—two were delivered to the Lawrence Livermore National Laboratory and one was delivered to NASA/Langley; CDC kept one for internal use. While CDC discovered the reasons for the STAR's relative ineffectiveness for scalar calculations, it stayed in the supercomputer race with the CDC 6600 and 7600. In effect, CDC produced the progenitor for the second generation.

Second Generation

Seymour Cray was the chief architect for CDC's supercomputers. He left CDC in 1972 to form Cray Research, Inc., where he designed and built Cray–1, the first vector-scalar machine. As with the first generation, the first markets were the government labs. However, Cray had a difficult time in selling its first computer to a government lab because of government regulations that required the machine be proven first. A unique six-month free trial arrangement with Los Alamos National Laboratory resulted in the purchase of the Cray–1; thus the company over-came some of the regulatory obstacles.

In the meantime, a number of CYBER models were developed for the second generation. CDC renewed its interest in supercomputers in the 1970s. It improved its STAR 100 components and added a scalar arithmetic unit. The early models were called CYBER 203 and were delivered to NASA/ Langley, the Navy's Fleet Numeric Oceanographic Center in Monterey, California, and the CDC Data Center in Minneapolis. Enhancements to CYBER 203 were made, and the improved machine became CYBER 205. In 1983, CDC decided to spin off its supercomputer effort because it believed that a small company could do a better job of developing a supercomputer than a large company. ETA Systems, Inc., is the spin-off. ETA later acquired the rights to market the CYBER 205 from CDC.

Denelcor (which folded in 1984) was the third U.S. firm to enter the second generation of the supercomputer race. Its machine was a nonvector, multiprocessor. It was a heterogeneous element processor or HEP.

Three Japanese firms entered the supercomputer race to try to meet the needs of their Japanese customers. The first Japanese company to announce entering the race was Fujitsu. Its second-generation machines are called VP 100 and VP 200 with peak speeds of 250 and 500 megaflops, respectively. Hitachi was the next with its S810/10 and S810/20 with 315 and 630 megaflops, respectively. NEC was the last Japanese company to enter the race. Its announced its SX–2 to be delivered in 1985 at 1.3 gigaflops.

Fujitsu currently owns 49 percent of Amdahl Corporation. It has announced that Amdahl will market the Fujitsu VP 100 and 200 vector computers to operate with IBM's System/370 software.

Third Generation

The third generation is in many respects a future system, meaning that it is still under development for later delivery. The companies in the second generation are all involved in the third generation. A brief listing of their machine titles and anticipated peak performance characteristics follows:

Company	Machine Designation	Anticipated Speeds (gigaflops)	Delivery Dates
Cray Research, Inc.,	Cray-2	2 - 3	1985
	Cray-3	10	1985-86
ETA Systems	ETA-10	10	1986
Denelcor	HP-2	2 - 3	1987

Some fourth-generation supercomputer programs are under way. One of these is the Defense Advanced Research Program Agency (DARPA) 1-trillion-flop machine being developed under the Strategic Computing Survivability Program. It is expected that $1 billion will be spent between 1984 and 1991 in four areas: extension of artificial intelligence, multiprocessor architecture, advances in VLSI, and rapid turnaround and fabrication of integrated circuits. DARPA's goals are substantially above Japan's MITI's national supercomputer program for a 10-gigaflop supercomputer in 1990.

Lawrence Livermore National Laboratory, under the auspices of the U.S. Navy, is developing the S–1 multiprocessor project with speeds up to 90–1,500 megaflops. At the high end of S–1 speed, it would be the fastest supercomputer currently being developed.

Some important implications and trends are evident from this brief review of the supercomputer generations:

1. In June 1985, the first of the third-generation machines, Cray–2, was delivered. Others will be delivered in the 1986 and 1987 timeframe.

2. The economic and scientific global competitive race for supercomputers is in the third generation of machines.

3. The U.S. government through DARPA and Livermore Laboratory is developing the fastest and largest of the third-generation machines. Their anticipated specifications are two orders of magnitude faster than CYBER 3, ETA–10, or MITI's supercomputer design specifications.

4. All dominant U.S. supercomputer companies can be considered emerging companies. None is in the Fortune 500 or dominant in the large mainframe electronic data-processing industry. None is vertically integrated as are the Japanese companies.

5. Supercomputers are at a critical junction. The second generation is being phased out as the third generation comes on stream. In the past, such generational transitions caused financial difficulties. The problems are likely to occur again because of manufacturing start-up costs for the third generation and shut-down expenses connected with the second generation.

There is no question that the U.S. government has been and still is the major investigator for supercomputer research and development and a key user of computers. One can question why the supercomputer industry has taken so long to develop. The answer is that to date larger-scale computing markets were outside the purview of the supercomputer. When the early supercomputers were developed for the government, their computing power and other abilities could be marketed or surpassed by large mainframe computers that manufacturers could develop to meet business and industry applications. As a result, mainframes became the core of the U.S. computing industry. Furthermore, they were very profitable. The rapid pace of developments of large-scale processing resulted in newer generations of machines for data processing with large data bases and an ever-growing market. Today the mainframe computer business is over $17 billion. Although over 106 supercomputer systems have been sold and delivered to date, the expanding market for these machines still promises $500–600 million a year for manufacturers.

At the moment, the supercomputer industry appears to be formless and highly unstructured. To put our arms around the state of the supercomputer industry, we must pull together various fragments that make up the industry. This includes R&D, supercomputer manufacturers, support infrastructure, software, and communication networks.

Research and Development

Research and development for supercomputers is more indirect than directed. Over one-third of the directed R&D for the supercomputers during the next decade will be funded by the government. The DARPA Strategic Computing Survivability Program research and development, while being conducted in four different areas, still has a 1-Tera-flop supercomputer to be developed as an end objective. The Livermore S–1 also has a specified supercomputer as a major part of its program. The other major supercomputer R&D programs are by the major manufacturers who are developing their supercomputers for the third generation—namely, Cray and ETA.

On the other hand, there is a larger amount of indirect R&D support that can be utilized for supercomputer R&D. These efforts are highly diffused and fragmented. Table 2.1 shows the IC^2 Institute's recent estimate that over $8 billion will be invested in both direct and indirect R&D for supercomputers.

The total funding for supercomputer and related research is almost as much as NASA's projection for the manned space station. It is a significant amount. The estimated funding is coming from three major sectors. We have projected federal government funding at approximately $3.4 billion with the dominant portion coming from the Department of Defense. The private sector is estimated to provide direct support of over $2.5 billion. The dominant portion is forecasted to come from the three primary manufacturers. Cooperative R&D support is composed of state governments, private corporations, and universities. The dominant portion would be private corporate research program support for the Microelectronics and Computer Technology Corporation (MCC).

R&D for the third generation of supercomputers is different from the second generation. The second generation was in many respects an outgrowth of the things CDC and Cray learned from their involvements

Table 2.1
Estimated Direct and Indirect Support for Supercomputer Developments (in the next ten years; in millions of dollars)

1.	Federal Agencies		
	A.	DOD	Strategic Computing Survivability Program	$ 1,000
			Strategic Defense Initiatives, Robotics and Artificial Intelligence	1,200
	B.	NSF	Supercomputer University Centers	200
			Communication Network	5
	C.	DOE		200
	D.	NASA	Space Station, Automation, Robotics and Artificial Intelligence	600
	E.	NIH	Medical Information Systems, Biotechnology Knowledge Bases	100
	F.	Other		45
				$ 3,350
2.	Supercomputer Companies		
	A.	Primary Manufacturers		$ 1,500
	B.	Secondary Firms and New Start-Ups		1,000
				$ 2,500
3.	Cooperative R&D		
	A.	MCC - Consortium		$ 1,000
	B.	Semiconductor Research Corporation		100
	C.	Stanford University Center for Integrated Systems		300
	D.	Massachusetts Microelectronic Center		100
	E.	California Microelectronics Innovation & Computer Science Program		100
	F.	Minnesota Microelectronics & Information Science Center		100
	G.	North Carolina - Microelectronics Center		100
	H.	Florida State Supercomputer Computations Research Institute		100
	I.	NSF University Supercomputers Centers - Matches		300
	J.	Washington-VLSI Technology Consortium		15
	K.	Indiana Computer Integrated Design, Manufacturing & Automation Center		20
	L.	Other		50
				$ 2,285
		TOTAL		$ 8,135

Source: All tables and figures in this chapter are from IC^2 Institute, The University of Texas at Austin, unless noted otherwise.

in the government-university-driven market for supercomputers. They invested their own funds and took all of the risk.

The third generation of supercomputers is still predominantly a private-sector development. Rather than being initiated by the larger U.S. computer companies as in the first generation, they are being developed by smaller computer companies. By computer company standards Cray and ETA Systems, Inc. are small companies. Cray is investing 20 percent of its sales revenue in R&D. ETA is still not independently financed. Each of the supercomputer companies is involved in investing $90 million or more of 1984 dollars to develop its third generation of supercomputers. Such R&D investments will be a continuing need for each succeeding generation.

These companies are relying on basic component research to be conducted through semiconductor company in-house research or on government-sponsored research programs. While most of such R&D funds will be from the federal agencies, primarily DOD, a number of states have begun to sponsor microelectronic research generally under cooperative research arrangements between businesses and universities. The Semiconductor Industry Association has also sponsored and supported centers of excellence in component research as well as selected research projects at a number of universities.

R&D for more productive and efficient means to design and develop the components to design the architecture for the supercomputer, to manufacture and test the supercomputer, and to develop operational and application software is also being conducted across a large number of independent federal and state government agencies, manufacturers, and various cooperative research programs and projects.

The largest cooperative R&D program is that sponsored by the 21 companies comprising MCC at Austin, Texas. None of the three supercomputer manufacturers are among the sponsors to whom research and development results are released prior to general licensing three years after their availability to members of the consortium.

The establishment of the NSF University Supercomputer Centers has involved two of the three supercomputer companies—Cray, ETA, and IBM. These recently created centers are utilizing these machines:

University Center	Manufacturer's Model
University of California at San Diego	Cray XMP
University of Illinois - a new center for supercomputer research and development	Cray XMP
Cornell University--Center for Theory and Simulation in Science & Engineering	IBM 3084QX and FPS 164 and 264 scientific processors
Princeton University--John Von Neumann Center	CDC CYBER 205 to be subsequently upgraded to ETA-10

These university centers and others at Colorado State, Purdue, Florida State, and Georgia provide an R&D window for supercomputer manufacturers on current and future applications and their impact on supercomputer designs that lead to subsequent generations.

The supercomputer companies are directly involved with the Japanese supercomputer race in the marketplace. Both the U.S. response to the Japanese fifth-generation computer challenge, or ICOT Project as it is now called, and the supercomputer scientific and engineering race from an R&D perspective have been taken up by private companies forming MCC and by the DARPA-initiated Strategic Computing Survivability Program. To the best of our knowledge, the three current supercomputer companies are not directly involved with either program.

In addition, the United Kingdom and European Economic Community have launched five-year research programs in the area of supercomputers. These programs are generally referred to as the Alvey and ESPRIT programs.

Current Structure of the Emerging Supercomputer Industry

As an infant industry, the supercomputer industry is in an evolutionary stage. Since my last testimony before the House Science and Technology Committee on November 16, 1983, the industry has taken on a different nature. The industry has become more diversified by broadening its base with slower machines for computational purposes. It can be described as consisting of

• supercomputer manufacturers,
• superminicomputer manufacturers,
• add-ons to mainframes, and
• software, graphics, and other peripherals.

Supercomputer Manufacturers

Of the supercomputer manufacturers, there are now two smaller companies and one Fortune 500 company in the United States:

Manufacturer	Estimated Number of Systems Delivered As of June 1985
Cray Research, Inc.	61
ETA Systems, Inc.	39*
Denelcor**	6
Amdahl	0
	106

```
 * Includes CYBER 205 which is only manufactured by CDC
** Company dissolved in 1985
```

Both Cray and ETA are located in Minneapolis. Amdahl is located in Sunnyvale, California.

There are also three large diversified Japanese companies in the supercomputer field. They are:

Manufacturer	Estimated Number of Machines Delivered As of June 1985
Fujitsu	2
Hitachi	1
NEC	0
	3

There is no question that nondiversified firms dominate the supercomputer field in the United States. The opposite is the case in Japan. So far only Cray has reported profits, which for 1984 were recorded as $45 million. Denelcor's 1983 losses were $10.3 million on sales of $3.6 million.

Superminicomputers and Minisupercomputers

There are a small number of manufacturers who develop and sell computers that are not the fastest for scientific and engineering purposes that require the 1-megaflop range. These computers generally sell for under $500,000. Among the companies in this segment of the market are IBM, Digital Equipment Corporation (DEC), Scientific Computer Systems, Inc., Convex Computing Corporation, and Floating Point Systems.

Both IBM and DEC are traditional computer companies. IBM has in the past responded with models to service a segment of the scientific and engineering computation market. It has made a conscious management decision not to compete in the relatively small market for supercomputers. However, it has always had at least one of its core mainframe generations available for the computing market, for example, 370/195 and 3083. These machines sell in the multimillion dollar markets.

DEC on the other hand aggressively has pursued the scientific and engineering calculation market even though it has not entered the supercomputer markets. It has been selling superminicomputers that are typically 1/100 as fast as the Cray–2. Its minisupercomputers are the VAX 11/780 class, which has an operational speed of 1.1 million instructions per second. They sell for between $125,000 and $500,000 each. The VAX 11/780 class as a superminicomputer and VAX–1 supermicrocomputer (one with .36 million instructions per second and sells for $11,245 per machine) together have accounted for over 35,000 machines sold.

In the last few years a newer type of scientific-engineering computer has appeared, which can be classified as the minisupercomputer. These machines are the result of evaluating three approaches to make slower computers to solve supercomputer applications at a lower cost. The two approaches that were discarded were to try to make a superminicomputer run faster and to take an IBM mainframe and provide it with an add-on feature for vector instructions. A third approach was selected—to design a totally new architecture faster than the superminicomputer and less expensive with a new software or software compatible with Cray or ETA. One of the current minisupercomputer manufacturers is Scientific Computer Systems, Inc. (SCS) of Wilsonville, Oregon. It made its architecture compatible with Cray XMP after securing Cray Research permission. These minicomputers can become satellites for existing Cray users as well as reduce the Cray XMP loads by offloading to the SCS–40 which became available in 1986. The SCS–40 operational speed is specified to be one-fourth of the speed of the Cray–1 or 20–50 million instructions per second. The Cray XMP software is compatible with the SCS–40. The SCS–40 could be about 20–50 times the speed of VAX 11/780 for approximately the same price of the high-ended VAX, about $500,000.

A second minisupercomputer company is Convex Computing Corporation of Richardson, Texas. Its machine, the C–1, is a stand-alone machine that runs at about one-fourth the speed of Cray. However, its design is directed to compete with the superminicomputers and has exploited the development of AT&T's Unix operating system to compete with DEC's VAX. It has taken a broader market segment than just the current scientific-engineering computing market of the supercomputer manufacturers. It also hopes to sell its machine in the still-evolving computer-integrated-manufacturing market. The Convex C–1 sells for approximately $495,000.

Both Scientific Computer Systems and Convex Computing Corporation believe that their next generation could well be in the same speed range as the Cray–1, or 80–200 million instructions per second, and carry a sales price of approximately $1 million. Both Cray and ETA managements have decided not to compete in the low end of the market at this time. One can safely predict that other smaller companies and start-ups will join Scientific Computer and Convex for a share of the lower-end market.

Add-on Systems and Flexible Building Blocks

Another dimension of the supercomputer industry is not to utilize either a supercomputer or a minisupercomputer or superminicomputer. This approach is to add computational processors to existing computers. Their design, therefore, is a fine balance between the types of applications and kinds of mathematics involved. Some companies have utilized a strategy that allows them to design add-on systems processors to be attached to current mainframes or superminicomputers to handle two- or three-dimensional arrays of numbers as well as vectors. One of these companies is Floating Point Systems in Beaverton, Oregon. It has targeted both the VAX superminicomputers and the IBM mainframe to attach its FPS–264. When attached, the computational speed becomes as powerful as the Cray–1 or about 80–200 million instructions per second. The FPS–264 sells for $640,000. Another add-on system company is Teradata Corporation of Inglewood, California.

A number of companies are developing supercomputers by linking flexible building blocks or by using a multiprocessor approach. Two companies in this field are BBN Labs, a subsidiary of Bolt Beranek & Newman, Inc., of Cambridge, Massachusetts, and Hydra Computer Systems, a division of Encore Computer Corp. BBN has already installed 16 systems with a total of 227 processors at a cost of about $8,000 per processor.

Some experts did not consider multiprocessors as part of the supercomputer industry until more than 1,000 processors were involved. Thinking Machines Corporation of Cambridge, Massachusetts, currently has up and running a "connection machine" with 16,000 processors. It expects to deliver a first 64,000-processor machine to DARPA in the fall of 1985. The 16,000-processor model has established a benchmark of 250 million instructions per second or one-fourth faster than the Cray–1.

Another company in this class is Sequent Computer, Inc., with its system, Balance 800.

A number of R&D flexible machines are being developed at more than 50 universities and start-up companies. Flexible has developed a general-purpose computer called Flex/32 that performs parallel multicomputing while maintaining an "essentially unlimited expansion capability." Flex has made four installations as of May, 1985. One was installed at Trinity Technologies Corp. in Dallas, Texas, for process-control application. Another has been delivered to Structured Software Systems, Inc., of Irvine, California, for satellite tracking systems. A third went to Purdue University's Center for Parallel and Vector Computation for research purposes. The fourth was delivered to NASA/Langley Research Computation Group for numerical solutions to do with aerodynamics and structural mechanics. Table 2.2 summarizes selected deliverable computer types from microcomputers to supercomputers.

In summary, the supercomputer industry in many respects is becoming more than just the firms that produce the biggest and fastest parallel-processing machines. The fastest supercomputers are much like the larger mainframes for data processing. It is now evident that it is possible to build slower minisupercomputer machines that are both compatible as well as noncompatible with the faster supercomputer in terms of operations softwares. The ability and the need to network supercomputers as well as minisupercomputers is also evident. What is not clear is what the network protocols should be. Currently network protocols in terms of networking supercomputers is at its early beginnings. Networking for supercomputers has been undertaken with success by DARPA, DOE, and NASA. The next round of evaluation and implementation of networks will be funded by NSF. How these networks will be compatible to handle the several generations of Cray, CYBER 205, ETA, and Amdahl supercomputers is still too early to determine.

The other networking aspects are the ability to network minisupercomputers with the supercomputers. The needs are not difficult to state. They will be required to enhance individual scientific research as well as comparative research between scientists in different geographic locations. They will be needed to increase the effectiveness and efficiency of educating scientists, engineers, and other graduate professionals in business, economics, and other social sciences. The minisupercomputer when shared with the main supercomputer can help to encourage both research activities as well as the education and training of the next-generation students. The various networking needs, such as graphics and data bases, are not as far along for supercomputers for scientific research or education.

Another aspect of networking is that connected with computer inte-

Table 2.2
Speed and Price of Selected Microcomputers

Computer Type	Company	System Name	Speed in MIPS (*) Megaflops (+)	Price ($)
Microcomputer	Apple	Apple	.0005 *	$ 1,795
Supermicro-computer	Digital Equip. Corporation	Micro VAX 1	.36 *	11,245
Minicomputers	Digital Equip. Corporation	PDP-11/44	.4 *	29,950
Supermini-computers	Digital Equip. Corporation	VAX-11/780	1.1 *	125,000 to 500,000
Mainframe	IBM	IBM 3083	4.2 *	735,000 to 1,975,000
Minisuper-computers	Scientific Computer Systems, Inc., Wilsonville, Oregon	SCS-40	20-50 *	about 500,000
	Convex Computing Corporation, Richardson, Tex.	Convex C-1	60 *	495,000
Add-on Systems	Floating Point Systems, Beaverton, Ore.	FPS-264	27-67 *	640,000
Supercomputers	Cray Scientific Research	Cray 1	80-200 +	7,200,000
		Cray 2	800-1400 +	17,000,000
		Cray 3	10,000 +	?
	ETA Systems	CYBER 205	800 +	6,000,000 to 15,000,000
		ETA 10	10,000 +	9,000,000 to 20,000,000
	Amdahl	Vector 1100	267 +	7,700,000
		Vector 1200	533 +	10,700,000
Flexible Parallel Processors	Thinking Machines Corporation	connection machine	250-1000 *	?

grated manufacturing (CIM) concepts including VLSI manufacturing. While the supercomputer industry can play an important role in CIM, it is still in its early stages. The barriers are standardization of network, cost, lack of expertise at top management levels, and a shortage of systems engineers and other trained personnel to implement CIM. However, the fact that we are seeing the supercomputer industry emerging and evolving into minisupercomputers, flexible parallel computers, and add-on systems is encouraging. This indicates that there are more dimensions to the marketplace than governmental scientific-engineering computations, university scientific-engineering research and training, and selected industrial research and design.

Applications, Markets, and Trends

The applications for supercomputers are still those that I presented in my testimony of November 15, 1983. The following quotation from that testimony is appropriate:

The supercomputer is the single greatest impact on world communication, automated factories, health care delivery, biotechnology production, renewal of basic industry and heightened productivity of the service industry, including government. The real task is how to develop appropriately integrated policies, regulations and support mechanisms that extend the U.S. computer/communications industry. The commercialization of supercomputers for the global market is so tightly structured from scientific exploration to ultimate use, regeneration time is so short, investments so large and risk so great, that we cannot leave policy consideration to evolve accidently and independently as in the past. The computer/communications industry is our future source of large scale employment; i.e., one out of six. Table 2.3 emphasizes the importance of the supercomputer as a driver for the computer/communications industry.

The consequences of losing economic and scientific preeminence in the supercomputer industry are vast. The supercomputer is a central driver for the rapidly emerging worldwide computer/communications industry. It impacts communication developments, the renewal of basic industries, productivity increases and the development and expansion of new industries. It is essential in improving our educational structure, fulfilling critical manpower requirements and enhancing our industrial creativity and innovation. It is the seed for encouraging the emergence of a myriad of technology venture businesses in the context of a private enterprise system that has always been the unique American way to achieve and maintain U.S. economic and scientific preeminence.[4]

Applications for fourth- and later-generation supercomputers are primarily for research and education. The research markets are the government and industrial research laboratories and academic research. The research and education market has gained momentum particularly during the early 1980s. The stimulus in many respects came from accom-

Table 2.3
**Effects of Preeminence in Supercomputer Technology on the Computer-
Communications Industry**

Drivers	Area Affected

Industries developed and expanded by super- computer-related technologies	Robotics-industrial automation
	Computer-aided engineering
	  Computer-aided design
	  Computer-aided manufacturing
	  Computer-aided testing
	  Computer-aided quality control
	Biotechnology
	Office automation
	International financial services
	Service industry
	Computer industry
	Semiconductor industry

Revitalization of basic industry	Energy
	  Petroleum
	  Nuclear
	Steel
	Automobile
	Textile
	Chemicals

Institutions receiving significant productivity boosts	Government Defense Education Research
Communications technology enhancements	Subscribers equipment Telephone company equipment Overseas carriers Satellite carriers

Source: George Kozmetsky, Testimony on Supercomputers at Hearing before Committee
on Science and Technology, U.S. House of Representatives, November 15–16, 1983,
92–94.

plishments in the government and industrial laboratories' applications
for aircraft aerodynamics, three-dimensional modeling of oil reserves,
emergency shutdown of nuclear reactors, utility power-grid planning,
molecular analysis, structural analysis for automobiles, shipbuilding and
skyscrapers, computer-generated imagery for movies, medical diagnosis
and product design, computer-aided design manufacturing and testing,
and others. These successful applications are currently moving into business and industrial applications.

A number of other research and educational applications require the
capabilities of third-generation and succeeding generations of supercomputers. These applications are what Neil Lincoln of ETA Systems, Inc.,

defines as leading-edge efforts in science and engineering that cannot be solved by the currently available supercomputers. More simply put, there are applications that are at least a generation ahead of the third-generation supercomputer capacity and abilities. These applications include the needs of the physical scientist, engineering of large-scale programs and processes, management of large-scale projects and programs, long-term economic modeling at global, national, and local-community levels, and others. It is perhaps important to stress that the research and educational market has a critical need for supercomputers that are of the nth generations beyond those in development. These are machines required for research in knowledge processing or what the Japanese called the fifth generation. Knowledge-processing machines will provide the momentum for future research efforts to be transferred and diffused for applications in the service, business, and industrial markets. Ultimately, there is a need for intelligence processing that is beyond knowledge processing in that these machines will make it possible for new knowledge to be acquired from utilization of large-scale knowledge bases (yet to be organized, built, and maintained) through computer learning.

The applications in business and industry are for the aerospace, automobile, energy, and chemical industries. The firms in these industries can be characterized as large and capital-intensive. They have been using supercomputers for developing specific applications such as wing design, simulation of in-flight conditions, seismic data analysis, designing new synthetic materials, genetic engineering, and total factory automation evaluations. These industries are becoming the market for second- and especially third-generation supercomputers as well as for minisupercomputers, flexible parallel processors, and add-on systems.

The market for supercomputers, particularly since 1983, has been a surprise to manufacturers and others. The surprise is that it not only has exceeded the heretofore early projection for a maximum world market of 100 supercomputers but that Cray alone has sold over 100 actual processing machines. The total number of sales by systems (some may contain more than one central processor) for U.S. manufacturers exceeded 106 by June 1, 1985.

Of equal importance is that the market mix has shifted from government labs to more industrial and university sales. Currently, it is estimated that the government-lab markets are about 35 percent of the total market.

There is a need to segment the markets and applications for supercomputers for policy purposes. One segment is the research and education market. Another is the business and industrial applications market. Each segment's needs and applications are totally different. Yet they are interlinked and supportive of each other, particularly in developing the overall supercomputer industry, including minisupercomputers, add-on systems, flexible parallel computers, graphics, and software.

The research and educational applications and their markets are also expanding. Earlier applications and market demands were largely federal-government driven. The current shifting of NASA efforts from the Space Shuttle to the Space Station, of DOD to Strategic Defense Initiatives, of DOE from alternative energy research to nuclear defense and weapons are all major programs that will require supercomputers. The national physics supercollider program will also require a large investment in supercomputers.

One should applaud the NSF University Supercomputer Center Grants. These grants are important steps in helping to ensure the overall scientific preeminence of the United States, particularly in the areas of particle physics, biotechnology, aerodynamics, and weather monitoring and prediction. These grants, together with other U.S. university supercomputer centers at Georgia, Florida State, Purdue, and Colorado State and others who are not dependent on NSF and DOE, no doubt will provide an important thrust toward ensuring U.S. preeminence while pushing the frontiers of U.S. basic research. Their research will help to expand the supercomputer industry, particularly for minisupercomputers, graphics, artificial intelligence, CAD/CAM, flexible manufacturing, robotics, knowledge base systems, and others. These grants, as expressions of a national policy, can begin to build important institutional links and informational networks systems among

- academia and industry
- scientific laboratories and centers doing complementary work
- scientific centers that require interactive scientific-technological dissemination
- researchers in academia and industry
- federal government, universities, and corporations
- federal government, state government, universities, and corporations

These linkages are relatively new. We do not have much experience with them, yet we cannot let important policy considerations evolve accidentally. The University Supercomputer Centers are primarily established to ensure U.S. scientific preeminence as well as to educate scientists and engineers for supercomputer usage. They have yet to be linked for maximum market and application purposes.

These current university-based centers are generally involved with basic research. The next five to ten years will see the traditional conduct of scientific and engineering research transformed to more computer simulation with less experimentation than required in the past. These shifts are taking place for several reasons. First, the critical needs of our nation such as defense, energy, and health are increasingly more multidisciplinary. Second, the research is becoming more complex and costly. Third, the science disciplines are increasingly interrelated and overlapping when addressing critical needs for new technologies. Fourth, there is increasing recognition that newer technology developments that are funded by and with government sources have an implied commitment to their short-term transfer for the purpose funded as well as longer-term diffusion for other economic benefits and positive impacts. The transfer and diffusion of government-supported R&D, especially for supercomputers and related research, has not been available since the first generation. As stated earlier, the second generation was primarily funded by the private sector.

It is important to understand how transforming the way basic research is conducted can impact current policies. There is a need to reexamine the traditional approach to intellectual properties and their dissemination. The issues include more than ownership, which is complex enough given the expanding network of relationships and competing claims. They include ensuring the vitality of and need for the centers and developing adequate provisions for transfer and diffusion of their results.

In the past little attention was paid to how science was transformed into technology that was subsequently transferred for specific commercialization purposes and then diffused throughout all industries, regionally as well as for international trade. The general paradigm was that basic research innovations would be utilized for applied research and development and that their manufacture would naturally follow. Diffusion to other uses and industries would occur when R&D results were both economical and better understood in general. The utilization of technology as a resource was perceived as an individual institution's responsibility. Economic developments flowed from this process because of individual firms' ingenuity and their entrepreneurial spirit. Targeting may have been a Japanese national policy, but in the United States, market opportunities at home and abroad seemed sufficient for economic growth and diversification. It was expected that all regions of the United States would in time enjoy the benefits of this paradigm in which new innovations from research were followed naturally by timely developments, commercialization, and diffusion.

The current experimental and collaborative uses of supercomputer centers are a break in the traditional way of doing research and perhaps subsequent commercialization. The major breaks are in the following two areas:

1. Financing is a collaborative effort. The other financial supporters are (a) companies that develop computers, (b) users of supercomputers and research

results, (c) state governments, and (d) individual universities comprising the consortium.

2. The research results are expected to be more basic research than proprietary products for industry. Both software and newer supercomputer designs are expected to emerge from the various centers and their participants.

There does not seem to be a built-in mechanism for self-sustaining the centers. If they are predominately doing basic research or advanced research for complex, multidisciplinary research, their end products are more publications of research results and less tangible goods and services. Publications and electronic transfer of research results may be important educational frontiers, but they may not be sufficient output for the centers' continued self-support. The replacement of currently selected supercomputers will continue to be capital-intensive. Maintaining and expanding software will be expensive, at least one-third of the cost of the supercomputer itself. Long-term funding will be a problem. What sources of research support for the centers will become available? Will they continue to depend on cooperative support?

The cooperatively funded supercomputer centers are primarily newer institutional developments of federal government, state government, business, and universities. Today the new institutional developments emerging around the supercomputer centers focus on maintaining U.S. preeminence in scientific research. Their role and scope as a policy matter at present is limited to five years of federal support. No built-in mechanisms exist to support either these centers in the long run or their longer-term roles for research, economic growth, employment generation, and economic diversification. Assuring that these centers do not become anachronistic institutions to be saved is a major policy matter. This changes the role of the federal government, especially in cooperative R&D and its subsequent commercialization for markets and applications for both the supercomputer industry and those upon which they impact.

In my opinion the NSF-sponsored University Supercomputer Centers, as well as federal government laboratories, are the leaders in trying to implement an emerging policy; that is, R&D that is funded by and with government and other institutional funds has an explicit commitment to technology transfer and diffusion. Federal government laboratories are in the forefront of supercomputer research and application. They are now seeking ways to transfer technology and ensure its subsequent diffusion. The IC^2 Institute at The University of Texas at Austin has been in close touch with Los Alamos National Laboratory to assist in the process of implementing effective technology-transfer mechanisms. The Los Alamos Laboratory has provided a number of workable newer institutional arrangements. Among these are arrangements for meeting places for people with common research interests such as the Center for Material Science, the Center for Non-Linear Studies, and a branch of the University of California's Institute for Geophysics and Planetary Physics. This also includes workshops for those in interested industrial firms and arranging for joint projects with other nations' participation. Dr. Donald M. Kerr, director of Los Alamos National Laboratory, has summarized the needs for effective transfer and diffusion as follows:

Finally, we try to recognize that both our very applied programs as well as our technology development programs have a need for people who are not purely disciplinary in nature, but in fact can play the new role of integrator, mediator, catalyst, or translator, whose technical breadth is required to work on these multidisciplinary activities, whose depth of knowledge has to be sufficient for them to be effective. We recognize that such people are more often found in the work place and trained on the job than produced through a traditional scientific or engineering education.

We see a long-term need to find a mechanism to encourage these people; to identify them, to give them the opportunities to develop their capabilities as they move from small to larger programs, both in relatively basic research as well as in the more applied development activities. This we think, coupled with the way in which science is increasingly done, taking advantage of large-scale computer simulation, will provide for effective and efficient research and development in the future on a timescale commensurate with national needs. The National Laboratories can adjust to meet the challenge I stated at the beginning, but it requires more effort than in the past, more willingness to experiment with new structures and processes, and more willingness to reach across boundaries to create partnerships with industry, financial institutions, and universities. The goal is to continually renew a creative anachronism to focus on multidisciplinary problem solving, enhance productivity, and create a pool of talent for the future.[5]

Transfer and diffusion require more than government laboratories or a university center's efforts. There must be institutional developments at state and local government levels and in the private infrastructure that assure such transfer and diffusion in meeting market needs and demands in a timely manner with fair distribution to all regions of the United States. There is a need for institutions to provide seed capital for economic utilization of the centers' research outputs for commercialization that cannot be the responsibility of the government laboratories and universities. Innovation in the United States is primarily a small and medium-sized company phenomenon. Currently, supercomputers are exploited primarily by the larger companies. This is true at present for the firms in the aircraft, automobile, and chemical industries. The stimulation of innovation through supercomputers for small companies today is a decided gap in policy formulation.

Current market trends for annual worldwide supercomputer sales indicate that sales on a units-sold basis are on track according to IC²

Institute projections in November 1983 for the U.S. House of Representatives Committee on Science and Technology. Figure 2.1 shows the early projections. The projections for 1984 for annual supercomputer sales in billions of dollars are higher than actual sales recorded by the industry by about 25 percent or $100 million. The projections did not take into account lease sales as such but projected them as units sold. The November 1983 projection scenarios are shown in Figure 2.2. The projections for supercomputer sales shown in Figure 2.2 are for the conservative scenario, which has been adjusted to account for sales from minisupercomputers, add-ons, other peripherals, and software.

Trends in the supercomputer industry can be summarized as follows:

1. There is an increasing demand for supercomputers that was unforeseen as recently as 1980. Demands are arising for federal government needs, expanding industrial and business markets (domestic and foreign), and universities (domestic and foreign).

2. All four U.S. supercomputer manufacturers are reporting a relatively robust firm backlog.

3. The smaller supercomputer manufacturers are finding a developing market for their products in education and research labs, as well as business and industry. There are strong indications that their next generation in development will broaden the number of applications and their markets.

4. Software firms for supercomputers are expanding and beginning to deal with emerging applications.

5. University Supercomputer Centers are focusing on advancing research and education in supercomputers.

6. University Supercomputer Centers, particularly consortiums at Princeton and the University of California at San Diego, can advance the communication networks between the collaborative supercomputer users. These advanced networks can be three to eight years ahead of the networking of large-scale, data processing computers.

7. The supercomputer network can and should avoid many of the problems that have beset the local area and long distance network problems for data processing and office automation.

8. There will be an increasing number of collaborative ties with the aim being to transfer and diffuse technology. These ties will include state governments, federal government, large and small companies, and universities.

9. International competition will become more fierce, especially from the Japanese in the next year. It is still too early to predict the competitive impacts from the ESPRIT program of the European Common Market and the United Kingdom's Alvey Program.

Figure 2.1
Annual World Supercomputer Sales, Projections for 1984 to 1993

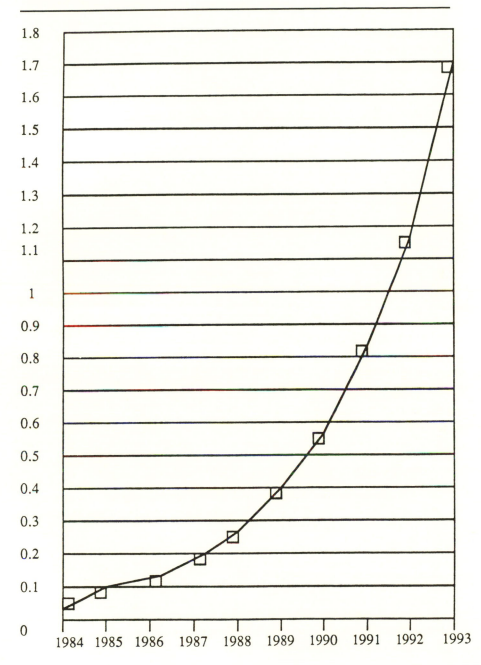

Figure 2.2
Annual World Supercomputer Sales, Projected Revenues, 1984 to 1993

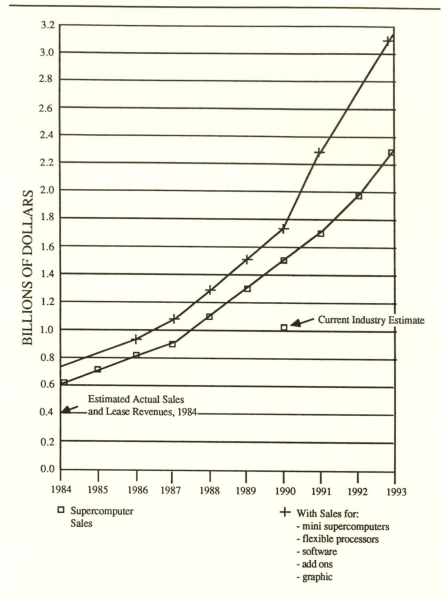

DISCUSSION OF THE ISSUES

1. *How will supercomputer R&D continue to be performed and funded in the future? What are the changing patterns?*

For initial supercomputers and the first generation of supercomputers,

R&D was funded by the industry and government. The first customers for the deliverable second-generation supercomputers were government labs. For the third generation of U.S. supercomputers, R&D is being funded by industry. This is the generation about which the Japanese announced they would be ready to enter into worldwide competition by 1990.[6] This program of R&D is funded at $400–500 million jointly by the Japanese government and six major, diversified Japanese computer firms.

The first known customers for U.S. deliverable third-generation machines are government labs and university supercomputer centers funded jointly by the federal government, state government, and computer manufacturers.

For the fourth generation of supercomputers, R&D is funded by the federal government. They are being developed by DARPA and the Livermore National Laboratory. The fourth-generation U.S. specifications are substantially ahead of the current Japanese 1990 goals and are in many respects a way of meeting the Japanese challenge. To the best of our knowledge, none of the four supercomputer manufacturers for the second- and third-generation supercomputers is actively involved with these R&D efforts.

Both funding and performing R&D for future supercomputers will be conducted under a number of different patterns. Among these are:

- Continue R&D with nongovernment funds.

- Enter into joint ventures with other companies, including foreign firms, licensing newer technologies.

- Participate in cooperative research and nonprofit consortiums like MCC for advanced research for enabling technologies to be selected for implementation by each participant.

- Enter into R&D contracts with federal agencies for a specific supercomputer or enabling technologies that can be used later for developing commercial supercomputers.

- Sell equity to other corporations that also license for their purposes the technologies being developed for supercomputer hardware and software.

- Utilize R&D partnerships and other traditional capital venture processes.

- Have government labs and/or universities buy or lease early models of next-generation supercomputers for use as beta sites.

- Establish collaborative links with research being performed by government laboratories and universities. These links can be to utilize relevant intellectual property funded by others such as the federal government or state governments or directly funded by gifts or business donations. Other collaborative linkages would be to sponsor university research for commercialization development and then retain all of the commercial rights.

2. *What kinds of complex applications and needs will drive the advanced developments of supercomputers?*

In the past a number of major computer manufacturers had dropped out of the supercomputer industry for a number of reasons. The primary one was that the market was not established or as profitable as that for data processing and automated office. The challenge from the Japanese was a major impetus for development of the U.S. fourth-generation supercomputers. Yet the second- and third-generation manufacturers have met the Japanese challenge. The emerging and still critical marketplace for the supercomputer world is being established by advanced government applications that are transferable into the supercomputer business and industrial market segments.

The current transfer and diffusion mechanisms need to be assessed to determine how to develop a market when there is not a replacement or when the substitution marketplace is required. Building a market using newer technology generally requires an infrastructure of specific needs that may not exist or be anticipated. In these respects, we know that there is a series of complex problems and special needs that even the fourth-generation supercomputers cannot solve. Among these are

- Large scientific research projects. Lawrence Livermore National Laboratory has already identified problems in nuclear weapons research that take 500–1,000 hours to solve on second-generation supercomputers. There are other such complex computational problems and research areas that will lead to Nobel prizes as well as newer markets and perhaps industries. In addition, knowledge processing and knowledge systems for large-scale programs for scientific research, for space defense and commercialization, for scaling up newer manufacturing processes, and for biotechnology are beyond the third-generation supercomputer capabilities.

- Advanced flexible manufacturing and processing computerized systems. These are beyond the third generation's capability computationally, as well as in terms of communication and networking among computers. There is no currently acceptable means of establishing standardization goals for more effective and efficient developments in a nondominant competitive environment. This is the near-term expandable market for the supercomputer industry, and it includes communications, robotics, CAD/CAM, CIM, graphics, and expert systems.

  The required data bases for these applications will take a long time to develop—between 5 and 12 years. There is no policy other than leaving it up to each user to use their discretion even for standards data. Many of the applications—including biotechnology, medical and health care, and financial services—are data-base dependent.

- Advanced security needs and applications. These areas continue as a supercomputer market. The needs for SDI, space, and overall national security programs provide a substantial market for both hardware and software de-

velopments. Government needs also include supercomputers for weather forecasting, advanced air-traffic control, and advanced data processing.

The critical question is to determine if it is possible to develop priorities
so that succeeding generations of supercomputers will not face a boom-
and-bust cycle or reactive crises and responses. It is essential that we
truly become anticipatory if we are to maintain U.S. economic and scientific preeminence.

3. *Can cooperative research reduce the development time and maintain the
U.S. competitive position in the global supercomputer marketplace?*

Recent institutional developments for cooperative research and development, including supercomputer applicability, have been initiated
primarily for economic growth, job creation, and diversification. These
collaborative efforts include industrial consortiums, software consortiums, university consortiums, and state-industry cooperative programs
with universities. Government labs have already been directed and are
actively pursuing collaborative efforts to transfer and diffuse their technological advances. Yet this is highly unstructured. How policies are
formulated will go a long way toward making and securing our nation's
and individual futures as well as the future of the supercomputer
industry.

Now I'd like to digress for a short historical overview to focus on this
issue. Prior to 1979, there was little evidence of technology venturing,
that is, collaborative or institutional developments for economic growth
and diversification. The prevalent attitude was a "go it alone" philosophy reflected in a variety of ways. The emphasis was on industrial
relocation rather than on building indigenous companies; separation of
institutional relationships, especially between universities and corporations; adversarial roles between government and business; and reactive rather than proactive policies both nationally and industrially to
meet international competition. We, at times, seemed to believe that the
rules of the game were set in concrete rather than subject to the dynamics
of an ever-changing global environment and to an economy that was
coupled to changing values. In addition there were the shifting roles of
the university in research and education.

As late as the mid–1970s, technology and its impacts were more threats
than opportunities with which to build a future. In 1978, total annual
venture capital was less than the then current one-day's loss of Amtrak
operations. Entrepreneurship was ignored as a force or driver. Technology transfer and diffusion were subjects for research and not a man-
date for commercialization of research and development. In the late
1970s, there was little doubt about U.S. leadership in high technology,
particularly in terms of electronics and its industrial and scientific markets. "Hi-tech" was unquestionably a major contributor to the nation's

trade balances. Then the loss of earnings, layoffs, and production cur-
tailments were not part of management's major concerns in high-tech
firms.

For much of the period from the 1950s to the 1980s, it was generally
assumed that scientific research would in one way or another transfer
into developments or technologies and subsequently be commercialized.
For much of this period, little attention was paid to how science was
transformed into technology, which was subsequently transferred for
specific commercialization purposes and then diffused throughout all
industries, regionally as well as for international trade. The general paradigm was that basic research innovations would be utilized for applied
research and development and that their manufacture would naturally
follow. Diffusion to other uses and industries would occur when R&D
results were both economical and better understood in general. The
utilization of technology as a resource was perceived as an individual
institution's responsibility. Economic developments flowed from this
process because of American ingenuity and our entrepreneurial spirit.
Targeting may have been a Japanese national policy in this period; but
for the United States, market opportunities at home and abroad seemed
sufficient for economic growth and diversification. It was expected that
all regions of the United States would in time enjoy the benefits of this
paradigm in which new innovations from research were followed naturally by timely developments, commercialization, and diffusion.

The current eight University Supercomputer Centers and those yet to
come are a good entry point for this issue. Government laboratories
have begun to transfer and diffuse technology. They have identified a
series of needs and especially for persons who can perform the role of
integrator, mediator, catalyst, or translator and whose depth of knowledge is both technically, financially, and managerially adequate for them
to be effective. These persons are not produced through traditional scientific or engineering or managerial education. Universities have just
begun to formulate policies and establish more effective organizational
structures to handle intellectual properties more economically. This is
not enough for what we are talking about under this issue. Professor
William B. Rouse has stated this issue as follows:

Despite the fact that basic research occasionally produces better mousetraps,
the applied world seldom visits the laboratory door. . . . The issues are many
and complex: they are more organizational in nature than technical.

Specifically, the university evaluation system should reward involvement with
real-world problems. Further, interdisciplinary cooperation should be vigorously
encouraged; current academic evaluation and reward systems, at best, only
tolerate such cooperation. Perhaps the best place for such changes to begin are
in professional schools such as medicine, law, and engineering where real-world
involvement would seem to be natural. Unfortunately, for these changes to

occur in engineering, for example, the tendency to emulate physics and mathematics will have to be reconsidered.

The university's relative immunity from "market forces" makes such changes feasible. For the same reason, organizational changes within the university are very slow. Thus, a short-term strategy for fostering technology transfer is also needed. It seems to me that any group of reasonable persons should be able to agree that new and/or vastly improved mechanisms are needed for transforming information in the research base into an appropriate form for the applications base.[7]

The Japanese seldom use academia in the actual targeted projects. They more often utilize business and government laboratory personnel. They also establish time frames for the project much like our "sunshine laws." They also accomplish transfer and diffusion when the researchers return to their firms or laboratories.

This is not necessarily what is required in the United States. Our social and work values are different. We do not utilize lifetime work principles of employment in either our firms or government laboratories. The newer supercomputer centers can modify and promulgate newer approaches to better enable their research to be more rapidly and efficiently incorporated in applied ways by firms. With this approach the past time span of 15 to 30 years to transfer technology into economic products and services can be substantially shortened. The utilization of intellectual property rights to better diffuse research results is also a good possibility. However, without explicit policy to address this issue, the traditional approaches will prevail.

4. *How can the results of cooperative research and supercomputer technology be transferred and diffused regionally?*

With government labs and university supercomputer centers as a prime focus we can better illustrate this issue. It applies to other cooperative research institutional developments that involve federal, state, and local governments, along with businesses and academic institutions in collaborative efforts. When national laboratories or centers are established, they produce, among other things, national technological resources. How to distribute these equally or fairly among the regions of our country is a newer issue. In the past and especially under full employment, this was not an important issue. Since 1980, various states have been concerned with unemployment, job creation, and the utilization of high technology. Over 150 initiatives have been taken over the past five years by the states.

The IC^2 Institute at The University of Texas at Austin has conducted a study to determine where the innovative developments are taking place. For these purposes, the major drivers used were federal R&D obligations, traditional venture capital, and selected company R&D expenses. The dominant states by ranking are shown in Table 2.4.

Table 2.4
State Rankings for Innovative Development

State	Dominant Federal R&D Obligations (fiscal year 1983)	Dominant Selected Company R&D Expenses (1983-84)	Traditional Venture Capital (1984)
California	1	3	1
New York	3	1	2
New Jersey	8	3	10
Massachusetts	4	5	3
Maryland	2		6
Texas	7		8
Illinois			5
Virginia	5		
Florida	9		
New Mexico	6		
District of Columbia	10		
Michigan		2	
Pennsylvania		5	4
Connecticut			7
Colorado			
Minnesota			9
Deleware		5	
Ohio		5	

Source: Compiled by the author from a study by Arthur D. Little.

At present, 17 states and the District of Columbia meet the innovative criteria. Four states are ranked in the top ten within each category: California, New York, New Jersey, and Massachusetts. These certainly are the first-tier states. The second tier or those in at least two of the categories' top ten would include two states: Maryland and Texas.

There are active developments taking place in restructuring firms and industries in terms of acquisitions and mergers. The dominant states, in rank order, are California, New York, Texas, Illinois, Florida, Pennsylvania, and New Jersey.

While a transformation is taking place, in terms of innovation, it has become evident that the present traditional paradigm that science and technology naturally evolve into commercialization that makes and secures a nation's future is inadequate. It does not adequately provide employment opportunities, mitigate layoffs, maintain a strong global competitive position, make economic security, and present growth opportunities across the board. Furthermore, the mechanism of allocating resources needs to focus more on flexibility and adaptability than on efficiency and effectiveness.

5. *How can supercomputer advances be transferred and diffused to small and medium-sized firms?*

Supercomputers as well as minisupercomputers are in many respects expensive and beyond the reach of many small and medium-sized firms. Firms of these sizes are already noted for their innovative abilities and as a major source for employment growth. As supercomputer developments promulgate, they will be among the first to utilize their advances, especially in design, manufacturing, and quality control. In some sense, if they could be placed in a position to utilize these results, they could help stem the flow of jobs outside the United States and at the same time increase other employment opportunities through new product development and increased productivity.

There has been little concern or organized effort to examine this need. This aspect could be achieved by incorporating some measures that utilized the university supercomputer center network system. This would help in the development of standardization for supercomputer hardware, peripherals, station terminals, graphics, and software. If done properly, it could help the growth of the emerging supercomputer industry.

CONCLUSION

The supercomputer industry is in its infancy. It is a promising period of growth. Its product lines are becoming diversified in terms of superminicomputers, software, and other peripherals.

Today it is an industry principally dominated by small, nondiversified supercomputer companies. They have managed to meet the Japanese challenge. They are successfully dominating the global supercomputer market. The long-term future for the supercomputer industry is still promising. It depends on how the user industries markets are developed. These markets are not necessarily replacement or mainly substitutions for current obsolete equipment, methods, and services. The markets are here now and can meet the visions for the new American dreams.

This industry is also pluralistic. It provides the key tools for basic research and advanced engineering. It will change the nature of higher education and may well provide the required thrust for longer-term means of financing academic research based on newer institutional arrangements under technology venturing.

Yet it is not a foregone conclusion that there are no additional policy issues that need to be attended to. A number of significant issues need the attention of the Science and Technology Committee. Chief among these is to assure that the University Supercomputer Centers are networked among the 30 university consortium members as well as the eight centers. This technological challenge, when met, can do much to keep this nation ahead of its foreign competition in Japan, the United Kingdom, and the European Economic Community, as well as the USSR and China. It is my prediction that we can gain three to eight years on them.

The networking, when coupled with the forthcoming U.S. third-generation supercomputers, can move the supercomputer industry from its infancy to the takeoff stage. It can also bring with it the CAD/CAM, Robotics, expert systems, and computerized flexible manufacturing industries into their adolescence and once more restore American manufacturing to world leadership.

Through the Science and Technology Committee, the U.S. House of Representatives can do much in formulating appropriate science and technology policy to assure that American small business and innovative entrepreneurs are assured of the availability of supercomputer technology at an earlier date than our traditional process of commercializing research has previously made possible.

ACKNOWLEDGMENTS

There are a number of friends and colleagues whose inputs and critiques have been especially insightful and helpful in the development of this chapter.

I would like to express my appreciation to Dr. Hans Mark, Chancellor of The University of Texas System. From The University of Texas at Austin, I would like to acknowledge Gerhard J. Fonken, Vice-President for Academic Affairs and Research, and Dr. James C. Browne, Chairman of the Computer Science Department. From the IC^2 Institute of The University of Texas at Austin, I would like to acknowledge the constructive suggestions of the following research fellows: Dr. Eugene B. Konecci, Dr. Ray W. Smilor, Professor W. W. Cooper, Dr. J. R. Kirkland, Michael D. Gill, Jr., and Dr. Stewart Nozette. I wish to thank Linda Teague for her invaluable assistance in preparing this manuscript and Patricia Roe and Ophelia Mallari for their research assistance and other support. Dean John Rouse and Dr. Ted Sparr of The University of Texas at Arlington and its Advanced Robotics Research Institute were especially helpful in providing insight for advanced supercomputer applications.

My appreciation also includes other academic colleagues: Professor C. V. Ramomoorthy of the University of California, Berkeley; Professor Edward Glaser, former director, Jennings Computation Center, Case Western Reserve University; Professor Robert M. Fano, Department of

Electrical Engineering, Massachusetts Institute of Technology; Frank Davidson, Systems Dynamics Group, Massachusetts Institute of Technology; Professor Herbert Simon, Carnegie-Mellon University; Dr. J. H. Poore, University of Tennessee; and Dean Robert Johnson, Florida State University.

NOTES

1. Kai Hwang, "Multiprocessor Supercomputers for Scientific/Engineering Applications," *Computer*, June, 1985, 57–73.

2. Sidney Fernbach, "Supercomputers—Past, Present, Prospects," *Future Generations Computer Systems*, July, 1984, 23–38.

3. Lloyd Thorndyke, Unpublished remarks to ETA Systems, Inc., Board of Directors, March, 1985.

4. Testimony on Supercomputers at Hearing before the Committee on Science and Technology, U.S. House of Representatives, November 15–16, 1983, 92–94.

5. D. M. Kerr, "Can Anachronistic Institutions Be Saved?" Paper delivered at the IC^2 Institute of The University of Texas Conference on Frontiers in Creative and Innovative Management, Miami, Florida, November, 1984.

6. Note that this is not the fifth-generation announcement by Japan. They have identified two target markets—one for supercomputers for scientific-engineering computations and the other for knowledge and intelligence processing. The first is the supercomputer program, and the second is the fifth-generation computer program.

7. W. B. Rouse, "On Better Mousetraps and Basic Research: Getting the Applied World to the Laboratory Door." IEEE Transactions on Systems, Man, and Cybernetics, January-February, 1985, 7–8.

3

PERSPECTIVE ON A NATIONAL SCIENCE NETWORK

J. H. POORE

The creation of a national science network is a major science policy issue. As such it is worthy of the attention of scientists and politicians alike. The goal of a national science network should be to create the technology base for complete and facile communication among scientists and their tools.

Collaboration is the basis for a large portion of scientific progress, and information exchange is the basic ingredient of collaboration. The tools of the scientific enterprise include word processors, bibliographic files, minicomputers, supercomputers, telescopes, and instruments too numerous to list. Every computer activity, every instrumentation activity, should be considered a component of what will ultimately be communication between people. Ideally, scientists geographically dispersed could share a common workbench, whether conducting experiments, writing papers, or using a shared resource such as a supercomputer. Ideally, scarce shared resources would be networked into a grid of availability. Scientist-to-scientist, laboratory-to-laboratory, campus-to-campus—whatever the affinity group, a properly devised science network should support interaction and communication at very high speeds.

The progress of the general scientific enterprise is suffering because it is unnecessarily difficult for scientists of like discipline and similar research interests to exchange data and images readily. Such activities are occurring today in a fashion that is dreadfully below both the state of the art and the capability of commercial technology. Nor are we realizing the full potential of our scientific resources such as supercomputers, accelerators, satellites, and other expensive instruments of scientific research. In installations that are not dedicated to network research, we see networking activities pursued in a parochial way that results in inefficient and ineffective use of scientific installations.

For example, several supercomputer facilities have or are developing networks for the purpose of linking their respective scientific constituents together. Some new centers are coming into being that will also grapple with this issue. Each is primarily concerned with attaching satellite persons and groups to its central purpose. Words are spoken indicating that careful attention will be paid to linking to other networks, but they are not convincing. Certainly this is an outrageous situation; but one cannot be outraged with any one of these uncatholic efforts because there is no clear alternative, no clear standard, no compelling leadership in national networking. In such circumstances, expediency is the driving concern.

Relatively few of the operational networks in the science community exist for the sake of network research and innovation. Relatively few of the supercomputers operating in the science community are instances of computer architecture or software research. Relatively few of the medical imaging devices are deployed for the purpose of research in medical imagery. Therefore relatively few such facilities need deviate from such standards as may have emerged or may be imposed.

The technical achievements of numerous national laboratories, industrial laboratories, and university programs are not denigrated by observing that each was constructed for a special purpose. Many technical networking achievements were of a research or demonstration nature, but proved so useful that they have been pressed into broader and prolonged use. For these innovations, criticism of their shortcomings in the broader interpretation is somewhat unfair. However, it is fair to criticize facilities and their organizations for clinging mindlessly to old ways simply because of local authorship or because of past investment. The issue transcends fairness and becomes a matter of national policy when new science facilities are put into place without full consideration of the general science constituency.

Policy concerns deal with the progress of the scientific enterprise as a whole, with the realization of the full potential of our scientific resources, and with the efficient and effective use of the public funds that sponsor the scientific enterprise. On all counts, the current situation in national networking is a policy issue.

Generalized networking is not a new idea. We have the example of ARPANET, which is somewhat exclusive, and the example of BITNET, which is very open. Most scientists are now using one or more networks and readily appreciate the value of a comprehensive network. Indeed, most scientists are already bothered by the necessity of dealing with multiple networks and the paucity of capability offered by each network. So, if the need is clear and examples are at hand, why are physical networks evolving around specific scientific facilities rather than around the concept of collaboration and communication?

We are currently committing two types of errors with respect to the issue of scientific networking. First, we are committing the intangible error of omission in that there is not yet a forum in which to explore the vision of total and instantaneous communication between any scientists and their instruments at any time without the constraining realities of programs, budgets, and technologies. Second, we are making the tangible error of headlong construction of networks of limited scope, limited function, and limited vision, without any serious effort of coordination toward interconnection and standardization.

In order to get the nation on the correct course for developing a national science network, it is necessary to establish a forum for discussion among the principal players. The Office of Science and Technology Policy is the coordination point for the various federal initiatives. Groups such as the Land Grant Colleges Association, Oak Ridge Associate Universities, Southern Universities Research Association, American Association for the Advancement of Science, Institute of Electrical and Electronics Engineering, and other learned societies play major roles in matters of this type and should have representation. Certainly, key industrial and university groups should be invited to participate. The purpose of such a forum is not to design a network, but to articulate the potential for a vast and powerful scientific facility.

The national science network must not be elitist. Using the word *campus* in a general sense to include industrial research facilities and national laboratories, the more campuses that are included in the national network, the greater will be the number of scientists in the network. Likewise, special facilities and institutions such as the National Library of Medicine, the National Bureau of Standards, and others must make their special repositories of information accessible through the national network. The national science network must be designed within the entire context of the community of scientists to be served.

We need a visionary concept for a national science network that presses the currently available technology. The concept should not be limited by jurisdictional considerations of the National Science Foundation or the Department of Energy or the National Institutes of Health. Nor should the concept be limited by disciplines of atmospheric science, or magnetic fusion energy, or aircraft design. Most important, the concept should not be limited by cost considerations or usage justification. At this juncture what is needed is a completely unfettered vision of how the nation's scientists might be given unlimited capacity to interchange data and images.

Only after the network is conceptualized should an engineering design be done. This is the proper point at which the realities of satellite availability, transponder capacity, antenna option, kilometers of installed optical fibers, protocols, costs, and other technical issues should be considered. This step must produce an engineering systems design that will pass scrutiny by competent communications experts, regardless of the cost.

I am confident that the resulting product—a national science network that is available to thousands of scientists through hundreds of universities—will compete effectively in terms of costs and benefits alongside other requests for research instrumentation that find their way to congressional attention. If a compelling design and a compelling story of the need and impact of such a network on U.S. scientific progress can be presented to Congress, then a lead agency will be designated and funding authorized and appropriated in due course.

PART II

PERSPECTIVES OF SUPERCOMPUTER MANUFACTURERS

INTRODUCTION

Supercomputer manufacturers in the United States have reached a critical stage in the development of their industry. Faced with a hiatus of aggressive federal support in the late 1970s and early 1980s, R&D lagged and technological advances came more slowly than might have been expected. Faced with a direct scientific challenge from Japan in the first part of the decade, the industry has been reduced to a comparative few companies that produce "supercomputer" hardware systems, and these companies find difficult hurdles as they seek appropriate financing to fund R&D, production, marketing, and sales. Accordingly, commentary from government, industrial users, and manufacturers call for collaborative ventures to ease the financial burdens and risk on manufacturers and their investors, while this vital industry becomes more securely established.

Concurrent to the business-financial challenge to manufacturers is the need to compete technologically. Today's supercomputer—the CYBER 205 and Cray XMP—will quickly be "ordinary" computers. Computer architecture evolved rapidly from serial processors to vector in the late 1970s and early 1980s, and vector machines are now being pressed by new advances in parallel processing technology. Hardware manufacturers and such users of high-speed computers are concerned that components and software lag significantly behind hardware development. Such a situation, of course, poses special difficulties for the development of the industry.

Part II presents the perspectives of major U.S. supercomputer manufacturers on these issues and their strategies for remaining competitive in emerging global markets.

4

SUPERCOMPUTER SYSTEMS MARKETS

LLOYD M. THORNDYKE

This chapter provides one vendor's viewpoint of the present and future markets for supercomputers.[1] Existing federal programs having impact on computer technology are evaluated and recommendations are made to address the fundamental underlying technologies. Some of the views are provocative and contrariant; nevertheless, we believe they are valid and should be expressed.

EMERGING COMMERCIAL MARKETS

Supercomputers are not an emerging commercial market. The current supercomputer market is dominated by industrial customers. ETA Systems commissioned a market study by a well-recognized market-research firm. Data from this study are presented here in table form showing the relative distribution of the current installed base of supercomputers and the forecasts of this base for the years 1990 and 1995.

	Market Share (%)		
Year	Government	Eduation	Industry
1985	38	17	45
1990	26	14	60
1995	23	14	63

These figures are all the more remarkable when one considers that the early supercomputer market in the 1960s and 1970s was dominated by the government segment with particular emphasis on the national laboratories and other leading-edge government users. The domestic higher-education market fluctuated with the fortunes of National Science Foundation funding, while the West European market maintained a consistent pace of a few installations annually. The only substantial, sustained industrial market during that early period was the petroleum industry, a long-time large user of high-performance computers.

Around 1980 the industrial segment of the market began to develop and accelerate to the point that today it outnumbers the government segment in terms of the installed base. At the same time serious competition began between Control Data Corporation and Cray Research. There seems to be clear consensus within government and industry that this trend toward increasing competition and industrial usage will continue, particularly as new applications are developed and put into use.

Our need therefore is not to recognize that the industrial market is about to emerge, but that it is a here-and-now market that will be accelerating its growth. The challenge we face as a nation is to be leaders—not followers—of the use of supercomputers to improve our economic competitiveness.

STRATEGIC DEFENSE INITIATIVE

It is appropriate to examine the words "Strategic Defense Initiative" in terms of supercomputers. It is difficult to underestimate the significance of the role that high-performance computing will play in the "star wars" scenarios—the implementation would be impossible without the availability of computing systems with supercomputer performance.

There is little question that supercomputers are "strategic" resources for the United States. Many, if not most, of the early systems were used effectively by the national laboratories in development of strategic weapons systems. Today the application of supercomputers has expanded and includes numerous uses of systems to develop and extend technology. The introduction of the supercomputer as an integral element of the design process of automobiles and increasingly denser integrated circuits are two civilian examples of this extension of the power of the supercomputer into areas of strategic economic importance. At ETA Systems we are using a CYBER 205 supercomputer to design our next system, the ETA–10. We will then use the ETA–10 to design its successor.

JAPANESE INROADS IN THE SUPERCOMPUTER MARKET

If we go back a few years and examine the supercomputers of the 1960s and 1970s, we will find that they were products with 100 percent

American technology content. All logic and memory components and the supporting peripheral subsystems were built in the United States. If you wanted to buy a supercomputer, you had to buy it from an American company.

Today, unfortunately, such is not the case and the situation is of more concern than is indicated by the mere fact that three Japanese companies have announced and started to deliver supercomputers, which are the direct outgrowth of the coordinated Japanese government and industrial thrusts, first into component development, then into general-purpose computers, and now into supercomputers and artificial intelligence.

There has been considerable publicity—most of it accurate—about the success of the Japanese in the memory-chip markets. Not only are they dominant in the "commodity-level" chips of low and moderate performance, but they are currently the sole suppliers of advanced chips that have the performance characteristics we need for supercomputers. The Japanese are also making serious inroads into the high-performance logic market, once the exclusive domain of U.S. companies, to the point that it is not unreasonable to visualize an American supercomputer designed around Japanese logic and storage components. We must not allow this perilous scenario to become reality. We must take the necessary corrective actions to make our semiconductor vendors more competitive.

FEDERAL COMPUTER-TECHNOLOGY PROGRAMS

There are several ongoing federal programs designed to advance computer technologies. Unfortunately, I do not see any of them directly addressing the key supercomputer strategic issues. Probably the best existing government program is the establishment of supercomputer research centers at key universities and educational consortia using funds being allocated through NSF. While the systems being purchased under this program contribute moderately to overall supercomputer sales, the more important contributions are the long-overdue investments in supercomputer software, applications research, and training of computer scientists.

The VHSIC program started by the Defense Department a number of years ago has had, at best, a trickle-down effect. Much of the effort, by necessity, was directed toward the unique requirements of the military market. I suspect that VHSIC is like all other programs of its type—there are trade-offs to be made and they will be made in direction of the main thrust of the project: meeting the military requirements.

Another new program has just started up in the Washington, D.C. area, calling for the establishment of a supercomputer research center to be administered by the National Security Agency and managed by the Institute for Defense Analysis. This program could be very helpful if its mission includes assisting the U.S. supercomputer vendors in the evaluation of technology risks. The Defense Advanced Research Projects Agency program has been under way for a short time with objectives directed toward artificial intelligence. Time will tell whether this effort will further our standing in strategic supercomputing.

It is not clear that the United States has established an effective program to protect our present supercomputer position from erosion. We have had several examples in the past in which U.S. companies have had cooperative computer technology and marketing arrangements with Japanese counterparts. These arrangements at their inception called for the U.S. companies to be donors of technology. In every case, we have seen a reversal of the roles—the U.S. companies are now importing Japanese products.

LESSONS FROM THE JAPANESE STORY

There are lessons to be learned from the Japanese success story, and it is instructive to examine our current position. First, we have established architectures that have been proven through customer usage. Second, our domestic user base is the largest and most experienced in the world. Third, we have the experience of having developed and delivered supercomputers longer than anyone else. This list expanded to an impressive size would seem to forecast a rosy future.

Our ability to compete in the industrial supercomputer market lies in getting back to the fundamentals and improving our technology infrastructure as the first step. The contrast in structure between the U.S. and Japanese supercomputer companies is dramatic. The Japanese supercomputers are the products of large, vertically integrated enterprises with the capability to produce all elements of the system, including the logic and storage components. On the other hand, the U.S. supercomputer suppliers are small by any standard of measurement.

In previous years, when the United States had a monopoly on computer technology, this was not a serious concern, since one could find cooperative domestic semiconductor vendors capable of delivering high-performance components. Today if we wish to buy the logic chips from domestic sources, we find ourselves in the semiconductor design business. If you want high-performance memory chips, you have no choice but to buy from Japanese sources. It has been our experience that the Japanese vendors do not make their advanced products available for use in the United States until they are in full production, assuring them of a substantial lead in the end products. This emphasizes again that the lack of domestic component sources must be of concern when we are dealing with an item of such strategic importance.

The place to initiate our line of strategic defense of supercomputer technology is at the very foundation of the business—the components. This is not an appropriate forum to debate the relative merits of one technology over another, but it is appropriate to identify some characteristics that most designers would agree are desirable. We need, first and foremost, a technology that will yield fast performance at the systems level. This implies not only fast switching times, but also high logic densities. This technology must be based on proven processes and supported by design tools and packaging techniques. For this approach to be successful, we need direct participation by the users of the technology—the supercomputer designers.

A successful supercomputer system does not consist solely of fast logic circuits. It also requires a large-capacity, high-performance memory subsystem, many fast input-output channels, and high-transfer-rate peripheral and communications subsystems and software to facilitate exploitation of the hardware features. Some of these elements are, I suppose, being pursued in various federal programs in one way or another, and that is the very crux of the situation. We have lots of fragmented programs that will yield us lots of fragments. What we need is a unified supercomputer program.

SUMMMARY

Despite current impressions to the contrary, we have already entered the stage that the industrial segment of the supercomputer market is dominant. The world is entering an era of significant supercomputer growth. The United States presently is in the forefront, but its leadership position is threatened by the impressive technological advances made by the Japanese. We need to come up with a focused supercomputer program that addresses the development of the fundamental technologies. Otherwise we risk loss of a key strategic element in both our defense and industrial sectors.

NOTE

1. This chapter also appears as Chapter 13 in Stewart Nozette and Robert Lawrence Kuhn, eds., *Commercializing SDI Technologies* (New York: Praeger Publishers, 1987).

5

SUPERCOMPUTERS: A POLICY OPPORTUNITY

F. BRETT BERLIN

In May 1983 I was interviewed by Cable News Network for a special news segment about supercomputers and our competition with the Japanese. Near the end of the two-hour interview, the reporter asked this particularly challenging question: "Putting aside the emotional issues associated with the United States' competition with Japan, and recognizing that the total supercomputer industry, worldwide, is less than one percent of the computer industry, why should anyone outside of Cray Research care what happens?" After some reflection, I suggested that American leadership in supercomputers was important for reasons of "national security, economic stability, and jobs."[1] One month later, John Carlson, executive vice-president of Cray Research and its chief financial officer, testified before the House Science and Technology Committee that there was a significant need for thoughtful policy in order to reap the benefits of the "coming of age" of this generation of supercomputer technology. He noted that "the world is on the verge of a major computing revolution as supercomputing is applied across a broad industrial and scientific base. The only question at this juncture is which nation will choose to lead that revolution."[2] In the two years following this testimony, this revolution—and its potential impact upon the defense and economic competitiveness of the United States—has been evident in words and actions of the Congress, the national security community, and both the research and the industrial science communities. Following the lead of the House Committee on Science and Technology, Congress has appropriated over $40 million per year for the last two years to guarantee availability of supercomputers to university researchers. The National Science Foundation has set up a special office to focus upon large-scale scientific computing programs. The Defense Department has initiated major technology initiatives aimed at very high speed computing. The Air Force, Navy, and Army have all begun major programs to obtain modern supercomputers for their in-house research facilities, and they are sponsoring research initiatives specifically focused upon development and application of very high speed computing to Air Force research and development needs. The National Cancer Institute of the National Institutes of Health has purchased a large supercomputer that will be dedicated to advanced cancer research. Major automobile manufacturers, oil companies, aerospace companies, and industrial research centers are using modern supercomputers for production "industrial science" applications. The chemical, biotechnology, and computer-animation industries are already positioned to become the next generation of major supercomputer users. Within the next ten years, it is likely that the applications of very high speed computing will have a major impact upon basic manufacturing, financial, and economic modeling, and even such "soft" applications as urban planning.

That the potential impact of supercomputing is broadly recognized is evident by the financial and technology trends of the supercomputing industry itself. In 1977, when Cray Research emerged as the preeminant supercomputer manufacturer, the total world market was estimated at around 80 supercomputer customers. Total Cray revenues were only $11.4 million from three customers. Cray's long-range plans called for growth to an eventual production capacity of a dozen or so systems per year. The Cray–1 system was essentially without head-to-head competition. At the end of 1985, the contrast was staggering. Cray revenues reached $272 million, 25 times the 1977 figure. Cray stock has grown in value by a factor of over 100 since its initial public offering in 1976; this stock was recently recognized as the stock with the highest ten-year growth average of any publicly traded stock. During 1986–1987 Cray Research will build more than 40 major systems comprising over 100 central processors; compared with total production of only three systems (one processor per system, at that time) per year in 1977. In addition to Cray Research, the industry has grown since 1982 into a fiercely competitive worldwide market, including as major potential players Control Data/ETA Systems (ETA Systems is currently a subsidiary of Control Data, which was carved out of CDC to allow a better focus upon high-speed system technology development and marketing), Fujitsu, Nippon Electric Corporation (NEC), and Hitachi. IBM, while not currently competing for the very high-end supercomputer market, can be expected to become a major player at the low end, particularly competing with Fujitsu and Hitachi. The supercomputer revolution has been considered so important to Japanese manufacturers, in fact, that this market has become one of Japan's carefully protected and "targeted" industries.

The technology-based revolution being ushered in by the broad application of supercomputers around the world has dramatic implications for the science and engineering community. One purpose of this chapter is to highlight some of these implications. Additionally, however, and perhaps more strategically important, there appear to be a number of public policy implications that need to be identified and studied. The second purpose of this chapter is to raise some of these issues. In order to accomplish these goals, I have divided my discussion into six sections. In the first, I establish what I mean by "supercomputer." The recent popularity of supercomputing has often resulted in confusion of supercomputing with other computing technologies such as artificial intelligence and very high speed minicomputers. The second and third sections identify several key technology and application trends, respectively, that could have major policy implications. The major impact that supercomputers are currently having upon the scientific and industrial world emanates from two types of breakthroughs: *system technology* and *application technology*. System technology, the subject of the second section, is technology that enables delivery of a system to the user, including hardware architecture, system software, and circuit technologies. Application technology, the subject of the third section, is technology that translates the power of the supercomputer system to a user in the form of a solution to the user's needs. The fourth and fifth sections deal with policy issues. Some current concerns that could be impediments to continued leadership in supercomputer design, manufacture, and application are identified and some key opportunities to leverage supercomputer technology developments to facilitate success in other areas are suggested. The sixth section concludes with several perspectives that I see as key to any supercomputer-related policy initiatives.

WHAT IS A SUPERCOMPUTER?

Neil Lincoln, principal designer of the CDC CYBER 205 computer system and other recently announced products from ETA Systems, once defined a supercomputer as a computer system that is "only one generation behind" the needs of leading-edge science and engineering. Others have described supercomputers simply as the "most powerful" general-purpose computers available at any given point in time. Whatever the specific definition used, it is fundamental to understand a priori that the term is subjective and is a function of both time and technology. Many, for example, of today's personal computers are "supercomputers" when compared with the general-purpose computers that were available in 1955. In 1972, the CDC 7600 was a supercomputer; by 1979 the Cray–1 had become synonymous with the term. However, even the Cray–1, the system that established supercomputing as an important trend, will not be powerful enough when compared with the next generation of supercomputer systems still to be called a supercomputer.

Table 5.1
Current Leading Supercomputers by Year of Introduction

First Customer Installation[a]	Relative Performance Class[b]		
	High-End[c]	Mid-Range	Low-End
1985	Cray-2 Cray XMP/4 (16 MWd Syst)	Cray XMP/2 (8 MWd Syst)	
1984	Cray XP/4 (8 MWd Syst)	CDC CYBER 205 (4-Pipe Syst)	Cray XMP/1
1983			Fujitsu VP100/200 Hitachi S-810/820
1982		Cray XMP/2 (4-MWd Syst)	Cray-1M
1981			CDC CYBER 205
1980			Cray-1S
1976			Cray-1A

a. Year of first customer installation does not always coincide with the year of product announcement.
b. Relative performance refers to throughput, not simply the latest MFLOP rating.
c. High-end rating indicates systems which are currently the most powerful available (approximately 10x "low-end")

Note: Systems less powerful than the Cray–1A (i.e., STAR 100, CYBER 203, TI ASC) are not *current* supercomputers. For the purpose of this table, performance is measured in terms of demonstrated capability to perform a user's actual workload. Common measures of high-speed performance, such as MFLOPS (million floating point operations per second) often obfuscate the only issue of genuine importance to the customer: cost to get the required work completed.

• Year of first customer installation does not always coincide with the year of product announcement.
• Relative performance refers to throughput, not simply the latest MFLOP rating.
• High-end rating indicates systems which are currently the most powerful available (approximately 10 x "low-end").
• Systems that were the leading, "high-end" systems when they were first installed.
Source: All of the tables and figures in this chapter were compiled by the author.

This fact is illustrated by Table 5.1, which shows the current major supercomputers by year of commercial introduction, categorized by relative performance class.

Figure 5.1 illustrates how the definition of supercomputers "migrates" as technology develops and is fielded. "Technology Trend" (shaded

Figure 5.1
Types of Computers

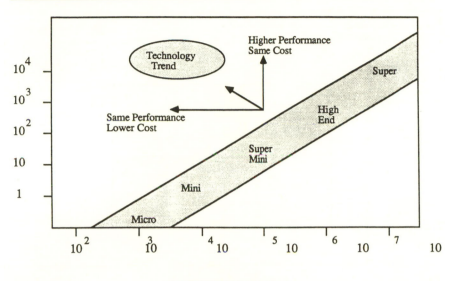

SYSTEM COST (Dollars)

area) is the "leading edge" direction for any class of general-purpose computing. In almost all cases special-purpose computers can, for a premium price, yield considerable performance gains. However, the cost effectiveness of these systems is poor enough that they are rarely applied outside of the domain for which they are designed. The Same Performance/Lower Cost direction results in application of high-speed technologies in such a manner that the cost can be reduced. Minicomputers were an early example of this type of migration; personal computers are the result of more recent development. Another recent development aimed in this direction is the new industry segment of systems called "minisupercomputers." While they certainly fall far short of even the low end of the current supercomputer range (by approximately a factor of ten, when measuring real workload performance rather than theoretical performance), these systems are specifically designed eventually to reach performance of near a Cray–1, at much less cost than the original Cray–1.

A proper understanding of the subjective nature of the word supercomputer is particularly important when making market projections or when identifying government policies important to ensuring continued U.S. leadership in the development and application of leading-edge general-purpose computing. For example, a number of projections have

Figure 5.2
The Potential Supercomputer Market

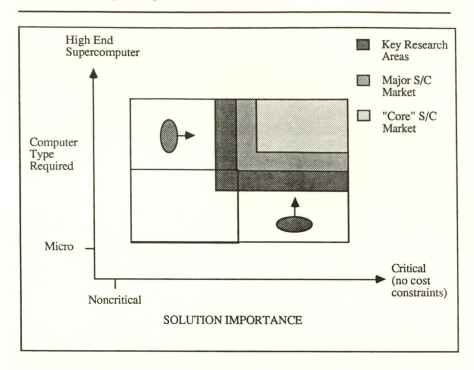

been made that indicate a market for hundreds, or even thousands, of supercomputers within ten years. One scientist recently predicted that supercomputers will be mass produced and sold at minicomputer or even personal computer prices by the mid–1990s (see Figure 5.2). The problem with such projections is that they forget that supercomputers of today will not be supercomputers in ten years any more than the CDC 7600 (or its modern equivalent, the CYBER 176) is considered a supercomputer today. What is undoubtedly true, however, is that there will be a very large new market for computers that perform in the Cray–1 class, but are available for less than 10 percent of the current Cray–1 equivalent cost. By 1990, such a goal is easily reachable, even without major unexpected breakthroughs. Given the strong market desire for such systems it is not surprising to see not only the large computer companies moving toward higher-speed scientific computing systems, but the emergence of a large number of companies offering very-high-speed minicomputers, many of which are specifically designed to handle applications that are today the preserve of large mainframe systems.

SYSTEM TECHNOLOGY TRENDS

The Cray–1A supercomputer system represents a number of key technology breakthroughs that resulted in a fundamentally new generation of high-speed computing. The supercomputer industry is now in the midst of another technology revolution that will render both the Cray–1A and the CDC CYBER 205 relics of past generations. This revolution hangs primarily upon six major trends.

Trend 1: *Utilization of Higher-Density, Leading-Edge, and Higher-Risk Circuit Technologies Rather Than Available Lower-Risk Technologies*

When Seymour Cray began building the Cray–1, in 1972, he chose for his basic circuit technology high-speed memory with 1,000 bits per chip. At that time, most other designers were relying on chip densities 10 or even 100 times as dense as those used by Cray. When Steve Chen designed the Cray XMP system, he built it around new, but stable, high-speed logic with density of 16 gates per chip. Others in the industry were talking about designing with slower, much more dense, experimental chips.

In 1983, when Cray announced the technology decisions for the Cray–3 system, he startled the computer community by committing to a design that would use semiconductor parts of gallium arsenide (GaAs), a highly experimental technology that was not yet being used by any commercial computer system. In order to support his decision, Cray Research built its own gallium arsenide development facility and set out to design not only a new generation of supercomputer, but a new generation of logic parts as well. These parts will not only be experimental almost until production quantities are required, but they will be over ten times as dense as circuits used in the Cray–2. Unlike previous efforts, the Cray–3 project includes sophisticated research in chip-level packaging, circuit structure, and fabrication technology. In essence, Cray's efforts are reinitiating the entrepreneurial process. Two years into these decisions, events have largely confirmed that Cray's decisions were well founded, but the initial decision was an incredibly high-risk direction to take.

Similarly, in his efforts to develop the next generation of supercomputers to follow the Cray XMP line, Chen has selected device technologies that may have to be built in-house because they are at the leading edge of high-speed silicon-based semiconductors. While staying within the silicon (Si) regime, Chen's design decisions are no less significant or high-risk because they required Cray Research to push the state of the art in components as well as in computer system design. At ETA Systems, the story is similar; designers are using VLSI experimental chip designs and, like Cray Research, are dependent on outside suppliers for both circuits and packaging technologies.

These decisions, and similar trends in other supercomputer development efforts, are significant for both technology and political reasons. On the technology side, the commitment to higher-density circuits is allowing the design of more compact systems with more complex functions. Parallel processing, for example, would not be feasible in high-speed systems without substantially more-dense circuits. More-dense technology also reduces the chip count per processor, thus reducing manufacturing complexity and increasing overall system reliability. Politically, the decisions of Cray have been significant because they illustrate a major weakness of the U.S. semiconductor industry vis-à-vis the foreign-merchant semiconductor industry. The U.S. supercomputer industry often must, in order to push its computer development as quickly as possible, make decisions that often have far-reaching implications in terms of technology transfer, production planning, and product marketing.

Trend 2: *Implementation of New Cooling Technology, Allowing Dramatic Reductions of Package Sizes and Overall System Space Requirements*

When the first Cray–2 was delivered in 1985, computing equal to the original Cray–1 was packaged in less space than the average executive desk. This dramatic size reduction was made possible by introduction of new efficient and maintainable coding system technologies. By 1992 supercomputing power will undoubtedly be available in a package as small as a programmer's work station. These developments, while caused more by the technology requirements than by design, could result in very-high-speed computers that can be transported in mobile environments or that can fit in operational systems, such as hospitals or factories.

Trend 3: *Implementation of Very Large Central and Auxiliary Random Access Memory*

When the Cray–2 was first shipped to a customer location, in August 1985, that single system had more central memory (256 million words or approximately two *billion* bytes) than all supercomputers ever shipped by Cray Research, *combined*. One writer claimed that this memory size was also larger than the cumulative value of all personal computers built in the United States in the previous year. Beyond impressive statistics is the fact that the tenfold increase in memory allows efficient realization of major engineering and scientific objectives that could dramatically affect the competitiveness of both U.S. and foreign industries that are able to profit from this new capability. The significance of this trend is further underscored by the fact that all major supercomputers in development today will have very large memory capability.

Trend 4: *Personalization of Supercomputing Resources*

The large computer was once viewed as a central computing resource that should only be used as a workhorse for very large problems. The idea of online access to supercomputers was considered by many to be a waste of valuable resources. However, as supercomputers have become more powerful, and as work-station technology has begun to provide the local terminal with considerable local power, the concept of delivering a large supercomputer system to a user "packaged" in the user's work station or personal computer has come of age. The supercomputer user of the future will have both hardware and software to support a full range of high-resolution graphics and other user-oriented interface tools, in addition to extensive high-speed networks designed to enhance the power and effectiveness of the local work station.

Trend 5: *Implementation of Parallel Architectures*

With the introduction of the Cray XMP–2 in 1982, the supercomputer community began its major turn toward parallelism. By 1987 it is unlikely that any very-high-speed computer system will be built that does not allow substantial parallel processing. Parallel processing stands to be a major, if not the most significant, force in computing for the next 15 years. In the last year alone, almost 100 new companies have been founded to pursue various parallel-processing-based products. While software and application-algorithm development are still major unresolved issues, the technology is now far enough along to warrant much thought over its true implications. Furthermore, if the implications are as dramatic as they appear today, there is a strong case for considerable, well-thought-through research into applications for this technology.

Trend 6: *Merging of Graphics, Logic ("intelligence"), and Numeric Processing*

Several years ago there was considerable concern, particularly within one defense organization, that artificial intelligence was the key to the future and that so-called numeric supercomputers were somewhat obsolete for all but a few applications. Since then, experience and research have indicated that rather than reducing the importance of conventional supercomputers, such special technology applications as artificial intelligence have made the development of supercomputers more important than ever. The significance of true, on-line three-dimensional graphics has had a similar effect, since the user community realizes that the key to both applications is the capability to move large amounts of data extremely quickly, and also that the supercomputer applications are becoming so complex that sophisticated graphics and artificial intelligence processing may be required to take advantage of the power of new supercomputer generations.

APPLICATION TECHNOLOGY TRENDS

Technology revolutions seem to occur in two stages. The first is the emergence of breakthroughs in the system technology that enable realization of new-problem solutions or dramatically improved versions of old-solution methods that may have had limited utility prior to the new technology. The second stage occurs when the system technology has been extant and used for sufficient time for the user community to understand how to use the capability as a lever for broad application enabling breakthroughs in other, previously unrelated problem areas. In one sense, the first stage is the stage of technology promise; the second stage—the subject of this section—is the stage of delivery. Six key trends are affecting this stage.

Trend 1: *True Three-Dimensional Modeling*

Perhaps the most important contribution of the Cray–1 supercomputer was that it allowed, for the first time, true three-dimensional modeling of a large number of simulation problems, within a feasible run time. This single breakthrough has literally transformed the thinking of many of the role of computer simulation in relation to experimentation. The significance of this trend is multifaceted. First of all, it allows the same problem analysis to be completed in much compressed time, allowing better use of the most valuable resource: people. For instance, in the aerospace design community, installation of a supercomputer reduced a process that took approximately a week to one that could be accomplished in a few hours. Second, and more significant, high-resolution modeling allows analysis of phenomena that are simply not testable experimentally, either because the resolution of the experiment is not sufficiently high or because the capability to be tested responds to another capability that has yet to be developed. This problem is evident, for example, in weapons testing, where one would like to test against projected rather than known threats.

The true implications of this trend are just beginning to be realized. When supercomputers are applied more broadly, many believe that the change will be fundamental to the basic process of science, design, engineering, and manufacturing.

Figure 5.3
Changing Environment for Supercomputer Applications

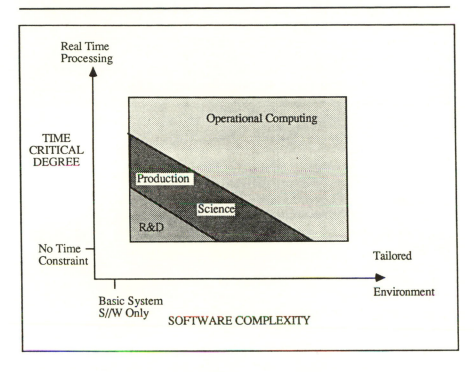

Trend 2: *Movement of Supercomputers from the R&D Environment to the Production Science and the Operational Computing Environments*

The technology trends identified in the previous section, coupled with the shift to modeling as a fundamental part of the development process, have resulted in the potential for an unusually exciting computing trend, as shown in Figure 5.3. Specifically, there is an opportunity to use the most powerful, leading-edge supercomputing as part of command and control networks, high-speed communications systems, or very complex medical care delivery systems. If this potential is realized, the implication for both defense and nondefense applications is substantial.

Trend 3: *Application of Supercomputers to Problems of General Information Processing Rather Than Strict Numeric Processing*

This trend has been alluded to, in part, in previous discussions of the use of supercomputers for graphics, artificial intelligence, and communications. The primary importance of this trend, which is just starting to surface as a potentially major direction, is that there are applications that cannot be done except when one can process entire pieces of information, such as images (medical, for example), text with meaning, or other nonstandard information representations common to people (users), but not common to computers. This year, both the Internal Revenue Service and the Social Security Administration were totally overwhelmed by their own data. It is becoming more and more evident that standard data processing is not enough. Another major economic application is the econometric modeling of various kinds required by the banking and financial services communities. The volume of data and interrelationship complexity is such that very sophisticated modeling may be required in the near future.

Image processing and graphic processing are already becoming major supercomputer applications that challenge even the largest supercomputers made today. Once again, the defense applications are substantial, as are the applications to critical areas such as cancer and other medical research.

Trend 4: *Utilization of Supercomputers for Software and Concept Development and Not Necessarily for Production Runs*

Initial indications show that complex software development can be completed more accurately for less than 10 percent of the development cost. If this experience continues to be repeated, future generations of supercomputers could be used for development of software for some of the complex but very-high-speed specialized parallel architectures being developed for special applications. Since over 90 percent of the cost of system development is generally conceded to be software costs, this trend alone could have major macroeconomic implications.

Trend 5: *Application of High-Speed Computers to Primary Problem Solving Rather Than Support Processing*

Trend 6: *Application of Supercomputers to Problems That Require System Installation in Nontraditional Environments*

These two trends are related to each other, as well as to Trend 2. The current generation of supercomputers has begun to make totally new applications feasible. The technology to make supercomputers has caused the systems to be reduced in size and has made possible, for the first time, the deployment of a supercomputer level of computing to operational problems that less powerful computers could simply not perform without human intervention, or that require on-site computing in locations where only smaller computers could be utilized in the past.

IMPEDIMENTS TO SUSTAINED LEADERSHIP

The U.S. computer industry has until recently been largely unchallenged in innovation, product development and manufacture, and in software and application development. This has been particularly true of the supercomputer industry, which until 1982 was considered the sole preserve of very-high-speed computing technology. Between 1982 and today, three Japanese computer manufacturers—Fujitsu, Hitachi, and NEC—introduced supercomputer products expected to be competitive with low-end and mid-range Cray products in production today as well as the first Control Data/ETA products expected for customer delivery in early 1987 or late 1986. While the U.S. supercomputing industry is still clearly leading Japan, there are a number of long-range policy and technology issues that could substantially affect the U.S. competitive position after 1988. Four key issues are either currently or will soon become active impediments to the growth of the American supercomputer industry. A second group of issues will, if handled with foresight, have a major positive effect on the U.S. competitive position.

Issue 1: *Supporting Technology Development Lag*

This issue is undoubtedly the one most discussed in studies of the supercomputer industry in the United States. The basic problem is very simple: Companies like Cray Research are not able, nor do they want, to integrate vertically to the extent that they can stay ahead of the requirements in every major supporting technology. However, as each generation of supercomputer is developed, the requirements of support technologies grow to the point that they hold back the development of the supercomputer itself. When this happens, the supercomputer company either has to wait—as Cray did with the Cray–2—or pick up the development effort in-house—as Cray has now done with semiconductors. The latter path is expensive and drains a major portion of internal R&D from the main mission, as illustrated by Figure 5.4. Obviously, if this trend continues unchecked it becomes very difficult for a smaller, focused company to move as quickly as it might otherwise. Fortunately, this problem is widely recognized as a key policy issue.

Issue 2: *Software, Algorithm, and Application Development Lag*

Supercomputers are developed primarily to solve user problems, rather than as a result of any national effort to be best. The best in-

Figure 5.4
Research and Development Budget Fragmentation, 1972–1985

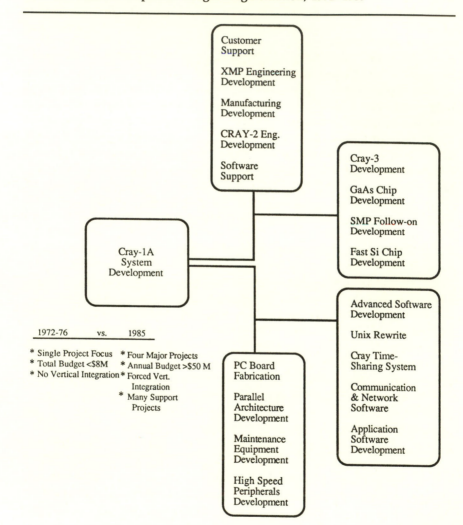

vestment in the future of U.S. supercomputer leadership is in efforts that can effectively move this development ahead of hardware, rather than behind it.

Issue 3: *Restrictive Procurement Policies and Actions*

The supercomputer, according to Bill Buzbee of Los Alamos National Laboratories, is like a telescope: Often one cannot predict what one will see until after the telescope is installed. If telescopes were controlled like most supercomputers, it is unlikely that we would have much astronomical innovation and discovery. This is particularly true in the federal government, where supercomputers are controlled as though they were simply large adding machines. If innovation is to occur and breakthroughs to be realized, then we must look for people with good hunches, rather than only for people who have problems that we already know how to do. The National Science Foundation University Supercomputer Centers are an excellent example of this problem. Already, the scientists with large "proper" problems are insisting that the resources be carefully distributed to those with the greatest "proven worth." However, there are a number of scientists who have applications that might be very exciting, but that have not been tried. We must have a policy that gives innovation some priority if we are to maximize the leverage available at a national facility. Figure 5.5 illustrates this principle. The traditional policy is to provide the time to applications in the darkly shaded area. The reasoning is simple: These applications have the money to pay. However, there is a large universe of other applications that, if the supercomputer can help, will yield very high dollar or mission leverage, but that may not have the budget dollars going into the process to take precedence over already proven applications. The problem of such resource allocation is not necessarily easy, but the rewards of overcoming the problem are high.

Issue 4: *Restricted International Market Access*

The U.S. supercomputer industry has become a major positive partner in research and development applications of computers throughout the United States and Europe and in the private sector in Japan. The U.S. industry has been carefully locked out of the Japanese public sector, and has much more restricted access to the commercial market than its Japanese competitors. While this lack of access has not had significant impact on the overall success of the industry, it is nonetheless of substantial strategic concern, particularly when the U.S. manufacturers are still clearly preeminent in product technology and capability.

Figure 5.5
Key Supercomputer Application Areas

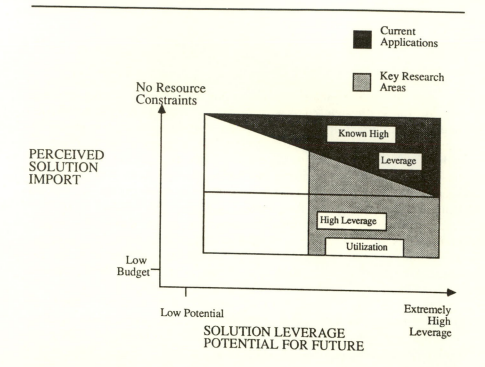

OPPORTUNITIES FOR INNOVATION

The previous sections of this chapter have focused upon the current state of supercomputing, taken from a policy point of view. The final two sections are focused on the future opportunities that are becoming evident for the next decade of supercomputing, assuming that the current rate of technology development continues and that the issues identified above are dealt with. In this section I identify major areas of public policy where supercomputing technology could have some major and fundamental positive effects. Five key opportunities made possible by supercomputer advances are these:

1. Foster a new environment for technology transfer, innovation, and entrepreneurial activity.
2. Model environmental and geophysical factors with sufficient accuracy to impact dramatically urban planning and environmental policy at all government levels.
3. Effectively test and evaluate products and systems early in the R&D cycle.

4. Dramatically advance the state of medical research and advanced diagnostic information processing for both treatment and diagnosis.

5. Accelerate the overall development of U.S. information-processing technology, particularly manufacturing technology.

Each of the five opportunity areas identified above is substantial enough to warrant a separate chapter. As one can see, each follows from the combination of the trends and issues identified above, coupled with the desire to allow supercomputing to be an exciting catalyst in the next decade of meeting human and societal needs. The first opportunity area deals with the need for an innovative society that can translate the massive results of our national research and development base into products and jobs. There are many ways being tried to ignite this process, but the most effective is to provide a central tool that becomes a "magnet" of sorts for development and, in the process, becomes the translation mechanism for transfer of technology and innovation.

The second, third, and fourth opportunity areas are some of the results that have already been made possible by the transfer of insight across the supercomputer to a key need area. Opportunity area 2 is the application of insight learned in such areas as oil exploration to the immense problems of toxic waste identification and cleanup, aquifer water supply problems, and other major potential issues that affect the average citizen at the local level, but for which there is no current solution.

Finally, the fifth opportunity area relates to the long-term and very-high-leverage issue of information, rather than numeric data processing. The area of manufacturing technology seems to me to be the most crucial, since our national industrial competitiveness is so firmly based in our leadership in manufacturing. It is clear to me that major resources and efforts need to be expended in this direction, and that supercomputers could accelerate the realization of breakthroughs by years and commensurate costs.

FINAL OBSERVATIONS

This chapter has focused upon trends, issues, and opportunities that the broad utilization of supercomputers have brought to center stage. They present us collectively with both the need and the opportunity for carefully thought-through high-leverage policy. This policy is not just at the federal level: The lack of federal dollars is not so much a reason supercomputers did not make it to many U.S. universities until the last few years as an excuse. The reason is that we often lack the vision to look ahead and to apply new technologies before they arrive, rather than after they begin to mature. The purpose of this chapter is to encourage that type of vision now. In order to implement that thought process, I

Figure 5.6
Supercomputer Policy: Key Players

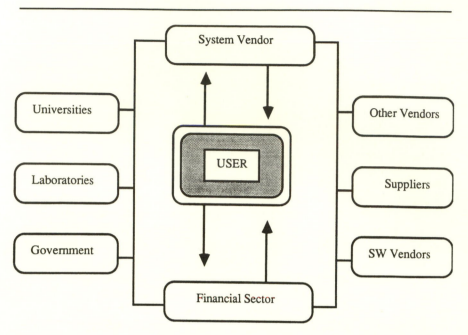

will close by suggesting three perspectives that have been underlying assumptions for my thinking on the subject:

1. General-purpose supercomputers exist because of and for users. All policy must be made with this key principle in mind. If the United States is to retain its supercomputer lead, it will do it because we have problems to solve and are willing to focus the resources necessary to solve those problems. Figure 5.6 is one model we might use.

2. Technology does not solve political problems. Rather, new technology may speed up solutions, alleviate pain while a solution is being sought, or may reduce the cost of solutions. Supercomputer technology, therefore, is a catalyst, but it does not obviate the need for careful policy.

3. Usage of supercomputers is the best way to push development of new generations. Many policy makers have proposed that the best way to ensure continued U.S. leadership is by a national program of R&D. We clearly need such a program for our long-term competitiveness. However, we also need to provide purpose to the research, a national program of application. Recent history has demonstrated that a nation that does not *apply* its technology solutions to its leading edge challenges soon slips in its competitiveness in

both the economic and political marketplace. I am reminded of what General Hap Arnold said when reflecting on the Manhattan Project, that we dare not muddle through the next twenty years the way we have the last twenty. I have worked with Theodore von Karman for the last twenty years and was sometimes scared by the knowledge he had that we weren't using. I don't want ever again to have the United States caught the way we were this time.

NOTES

1. The ideas expressed in this chapter are the sole responsibility of the author, and do not necessarily reflect the opinions or policy of Cray Research, Inc.

2. John F. Carlson, statement before the U.S. House of Representatives, Committee on Science and Technology, Subcommittee on Energy Development and Applications and Subcommittee on Energy Research and Production, June 14, 1983.

6

THE PARALLEL PROCESSING REVOLUTION

R. DAVID LOWRY

INTRODUCTION

In the past, the major contribution to increases in raw computer performance and price-performance ratios has come from the components used in building the systems. Projections now show that between now and the end of this century components alone will not support sustained growth at previous rates. Given an ever-increasing demand for greater performance, there has therefore been a major worldwide movement to develop new highly parallel architectures that can support the performance growth requirements.

Parallel and multiprocessor architectures based on microprocessors show great promise for the future. However, current microprocessors have architectures that do not efficiently support massively parallel systems. This will limit the upper performance level of such systems. When new microprocessors are created with architectural features that support high levels of parallelism, many systems will appear that will perform at supercomputer performance levels. It will require five to ten years before microprocessors of this type are generally available and in widespread use in commercial systems.

There are critical economic and defense needs for supercomputers in the next five to ten years that can only be built with proprietary designs. Bringing such products to market through efforts of the private sector are prohibitive in our current environment because of cost and risk factors. Cooperative ventures between commercial developers of these systems have been encumbered by antitrust legislation, and U.S. government efforts to foster supercomputer development have lacked a central focus.

Recognition of these factors and coordinated efforts by government, industrial users, vendors, and educational institutions will be required in order to maintain our leadership position in this critical industry.

EVOLUTION OF THE COMPUTER

Advances in computer architecture during the past 40 years can be divided into three major categories: architecture, software, and components.

Architecture

The architecture of most commercially available computers still closely follows the original von Neumann model. That is, the instruction processor executes a single instruction (mathematical, logical, or control operation) against a single data stream at each step in the computational process. Computers of this type are usually called serial processors and represent the largest number of installed systems worldwide.

Computers that can do more "work" in a single step of the computational process were commercially available by the mid–1970s. They came in two basic styles: vector processors, and array processors. Both of these types of architecture took advantage of the fact that latencies introduced in various parts of the system could be "hidden" by replicating certain hardware functions that performed such stereotypical actions as address computation and operand fetching, or math operations. Thus a single instruction stream was able to perform operations on multiple data streams concurrently. For the narrow range of computational activities that can take advantage of it, this architectural innovation can provide dramatic performance advantages over serial computers. Most supercomputers today utilize this type of architecture, and are therefore limited in applicability to a small fraction of the range of problems that need computers.

It is now widely believed that computers must exhibit an even higher level of parallelism than the vector and array processors have, in order to maximize performance. Both theoretical and commercially available machines now exist that can execute multiple instruction streams against multiple data streams concurrently in a single program or multiple programs. This type of system is usually called a parallel processor. The potential exists for this type of architecture to perform well against a wide range of computational problems, thus making available to more commercial users supercomputer performance like classical management-of-information-systems functions, computer-integrated manufacturing, and transaction processing. The additional economic pull generated by broader markets will eventually have the effect of stimulating investment by the private sector in new highly parallel computer systems.

To assure significant numbers of such architectures in the market will require new, highly focused incentive programs to combine resources of government, education, industrial users, and vendors.

Software

In the life-cycle cost of ownership for a computer system, the percentage expended for software has grown faster and now exceeds that of hardware. This trend is expected to continue for many years. Software advances in the last few decades have contributed more to ease of implementation than to performance. That is, new higher-level languages, standardization, and programming environments have had a larger effect on the bottom line for computer users than have the development of optimizing compilers and other performance improvements in the software.

The advent of a new generation of parallel computers can build on the software development methodology we have developed, and can use the program development environments now commonplace. However, the high-level languages and operating systems will have major internal changes, and the programming paradigm based on von Neumann architectures must be replaced. Both of these areas require a huge amount of research leading to the formulation of new standards. Major restructuring of the training process for programmers and computer scientists must begin in short order or we will have a more serious technical labor shortage than even our current bleak forecasts indicate.

The correct priority for the research and training is to start with the languages that will support massive parallelism and their associated debugging tools, and then to formalize operating system standards for interchange of programs and data among heterogeneous systems.

Components

The greatest contribution to improved raw performance and price-performance ratios to date has come from the components we have used to build our computing systems. The steep curve of improvement may now be flattening in five- to ten-year projections because of speed-of-light, geometry size, heat removal, and quantum-physics limitations.

Continuing the system-level performance improvements therefore has led to strategies that involve parallel processing. In fact, in today's supercomputers, performance is restricted more by the speed of light through the various conductors in a processor than by the switching speed of the components. Shortening the aggregate length of those paths directly affects throughput rates in the computer. Since the vast majority of signals that must be transmitted in a computer system happen within rather than between processors, the parallelism strategy implies that if a larger number of relatively smaller (therefore faster because of conductor lengths) processors can be applied to a problem, the entire system will have higher throughput.

To achieve supercomputer performance with components currently commercially available it is necessary to implement a considerable amount of circuitry used only to hide latencies caused by the fact that storage technology and communications technology are slower than processing technology. That is, the throughput capacity of computers is dominated by the transit time of data in programs to and from memory and between processors. A much "cleaner" and more-cost effective computer could be built if those technologies could be improved in performance regardless of architecture; research and development in this area could yield huge returns.

Availability of supercomputers that have broad applicability as do shared-memory, scalar, multiple instruction-stream, multiple data-stream (MIMD) computers offers the possibility of broadening the market for these machines to commercial ventures. In these environments, advanced component technologies that require exotic cooling techniques (e.g., immersion in cooling liquids or liquefied gasses) are not acceptable. Fortunately, such architectures do not stress component technologies beyond their predicted performance capabilities within the next few years.

However, even these commercially available components represent a relatively unattractive market to the merchant vendors since there are only a few supercomputer manufacturers, and their present volume is very small. Incentives to the semiconductor manufacturers to continue their interest in this area would represent a boost for the supercomputer industry.

CURRENT ENVIRONMENT

There are currently only two or three supercomputer development efforts in the United States funded by the private sector. The reason for this is the high cost and risk involved in pioneering the new architectures and packaging techniques involved in supercomputers, compounded by the need to develop software that operates on massively parallel computers.

The most recent supercomputer, and the only shared-memory, MIMD machine commercially available, is the HEP System from Denelcor, Inc. (which folded in 1984). From 1978 to 1984 Denelcor spent over $50 million producing the system. Government assistance was crucial in order to perfect the new architecture so that this amount of money could be raised from the private sector. The U.S. Army Ballistics Research Laboratory and NASA funded the research phases of the HEP System development because of the potential return such an architecture would have in solving problems crucial to their activities. Government funding amounted to approximately the price of one system, but the timing of this contribution was most important in producing the first such parallel architecture.

It is estimated that an equivalent effort today would cost more in the range of $150–200 million instead of $50 million. It is therefore highly unlikely that the private sector will fund enough such efforts to maintain a U.S. leadership role while producing sufficient competitive efforts to produce a viable marketplace with private-sector funding.

There is a commercial demand for both the vector or array-processor architectures and more general-purpose scalar MIMD machines that can perform in the supercomputer performance range. Modular multiprocessor machines can also extend the new architectures down the price-performance curve to lower the entry-level costs and dramatically broaden the market by extending availability to companies that have a legitimate need but who cannot support current supercomputer prices.

In addition to commercial demand, we are entering a phase of strategic defense requirements that will absolutely require radically new architectures. These include the Strategic Defense Initiative, advanced battlefield management, command-control-communications, and autonomous vehicles. New applications such as artificial intelligence will also pervade both the commercial and government markets and will require different and expensive new computer systems for high-end applications.

Nationalized efforts in Europe and Japan have been launched to challenge the current U.S. dominance in the computer industry. The huge effort to develop the next generation of supercomputers will have several effects. First, it will accelerate the development of required infrastructures that support the whole range of the computer industry's requirements. These include component and materials science, software development, computer architecture, and a broad range of support industries. The nations now engaging in this research recognize that their efforts would not just capture the relatively small supercomputer market but would also position them to participate in the over $40 billion computer market. The foreign nationalized efforts capitalize on their ability to cause cooperative efforts between industrial users, vendors, educational institutions, and government in ways that have been precluded in the United States by antitrust legislation and trade-secrets protection in U.S. industry.

In the past, the U.S. government-funded advanced computer research in universities has yielded very little in terms of spinning off commercial products. Special-interest groups, with the support of their local political representatives, have also had success in capturing funds targeted for the advance of computer science. This is partly because elected decision makers are ill prepared to evaluate the highly technical issues involved in choosing research centers that are both qualified and sensitive to the challenge of producing marketable computer advances.

REQUIREMENTS FOR A VIABLE SUPERCOMPUTER INDUSTRY

Financial Climate

Given the high cost and risk of R&D and commercial production of supercomputers, some assistance must come from government and the industrial-user community for promising ideas to be developed to the point where private-sector investment is feasible.

Strongest consideration should be given to architectures that have volume-market potential in both commercial and strategic defense markets as this will assure high private-sector participation.

Beyond the proof-of-concept phase the government must act as a "friendly customer" for early purchases of important new architectures. This does not mean buying computers just to keep the vendor in business. There are almost always requirements within the government for computers with advanced capabilities that are so crucial that additional monies that may need to be spent (e.g., to provide software or peripheral attachments not available with the early models) are well justified by the returns. It can, however, mean that exceptions could be made to regulations intended for general purchases of commercial computers. For instance, existing regulations require that the government not purchase computers that have not been supplied in significant commercial quantities. While these regulations generally have merit, neither the government nor commercial users would have supercomputers if they had not been successfully bypassed in the early days of supercomputers. It is time to formalize a procedure that protects the taxpayer's interests while fostering development of crucial economic and defense importance.

Given a viable and producible computer, the government can also provide direct monies to develop enhancements or software (e.g., new high-level languages or special interfaces that would not be feasible because of cost for the vendor). While this has occurred in the past, no general policy, enabling regulation, or centralized oversight group exists in the government to manage this process.

Infrastructure Development

The closest approach now in operation in the United States to the Japanese Fifth Generation Project is the Strategic Computing Program of the Defense Advanced Research Project Agency. Its statement of goals and its organization recognize that an infrastructure that includes universities, industry, vendors, and government is required for any self-sustaining effort toward next-generation computer technology. The flow of ideas from universities to industry and then to government users is crucial. The availability of a continual flow of trained scientists and engineers has slowed to a trickle in the supercomputer industry since these tools have not been available to qualified researchers.

The next generation of computers will already be available before we can fund and redirect university programs to provide a new class of trained users. The National Science Foundation has recently been funded to procure supercomputer resources for U.S. universities. However, even with this laudable program, the United States will lag behind other countries in per capita university availability of supercomputers. Furthermore, more investment is needed to fund course development and research projects that make practical use of the resources provided.

7

AMDAHL VECTOR PROCESSORS

WAYNE McINTYRE

The Amdahl vector processor systems were developed with three major design objectives: (1) high performance, (2) ease of use, and (3) high reliability. High performance is obtained through large-scale integration (LSI) technology, pipelined hardware architecture, and advanced software such as the automatic vectorizing compiler (FORTRAN 77/VP).

Ease of use is accomplished with the advanced vector processor application development system (VP/ADS) software and through the introduction of IBM compatibility to the vector processor environment. VP/ADS consists of the FORTRAN 77/VP compiler, a full-screen interactive debugger (DOCK/FORT 77) that allows the user to view and control program execution in real time; an Interactive Vectorizer, which prompts users to optimize and take application code; FORTUNE, a tuning tool to analyze FORTRAN program application; an extensive scientific subroutine library; and several other tuning and debugging aids.

IBM compatibility is provided through support of the S/370 instruction set, common operator commands and job control language, sharing of DASD between MVS front-ends and the vector processor, and support of standard IBM channel interfaces. The operating system on the vector processor is MVS/XA (a trademark of IBM Corporation).

The introduction of Amdahl vector processors has set new standards in scientific computing. Amdahl's focus on performance, compatibility, and reliability brings unique offerings to the supercomputing marketplace. By using the densest chips available, Amdahl vector processors achieve faster speeds at lower cost. Unique software tools allow a higher degree of vectorization, which improves performance. Since all models are compatible with System/370 architecture, user investment in hardware, software, job accounting, facilities, and training is preserved. This compatibility also ensures easy system integration, rapid application development and migration, and efficient operation. Reliability is an integral part of the Amdahl vector processor system design. By using proven components, innovative engineering packaging, and high-quality manufacturing techniques, the Amdahl vector processor is able to maintain a high level of reliability at very low operating costs.

The Amdahl vector processor models 500, 1100, 1200, and 1400 are Amdahl's largest scientific and engineering computers. These machines employ a multiple, concurrent pipeline architecture to obtain a peak performance of 142 MFLOPS for the model 500 to 1142 MFLOPS for the 1400. The 1200 and 1400 support up to 256-M bytes of main storage. Since peak performance is rarely sustained, the vector processors include several advanced vector processor features in order to achieve very high application performance for customer FORTRAN programs. This chapter describes the design approach, technologies, architecture, and software of the vector processors, with particular emphasis on the implementation of the advanced hardware and software features.

DESIGN APPROACH OF THE VECTOR PROCESSORS

Development of supercomputers has been motivated mainly by the need to perform such large-scale simulations of physical models as hydrodynamics, numerical weather prediction, and nuclear-energy research. As the extremely powerful computational capabilities of the supercomputers attracted attention from various branches of science and industry, however, uses spread to structural analysis, VLSI design, oil-reservoir simulations, nuclear-plant simulations, utilities, and quantum chemistry, just to name a few. This trend has resulted in customer requirements for supercomputers that can efficiently handle a wide range of users' programs.

In order to meet this requirement, more than 1,000 FORTRAN programs in the typical application areas were analyzed prior to designing the vector processor. The analysis identified the common characteristics encountered in scientific and engineering computations, which were then used in both the architecture and the compiler design. The detailed results of this study have been published.[1] We summarize here the conclusions from this study.

Although high-speed components, high degree of parallelism or pipelining, and large-sized main memory are the basic requirements for increasing the computational capabilities of a supercomputer, the following four advanced features in architecture are equally important for such a machine to be versatile enough for its wide range of intended applications: (1) efficient processing of DO loops that contain IF statements, (2) powerful vector editing capabilities, (3) efficient utilization of the vector registers, and (4) highly concurrent vector-vector and scalar-vector operations. These advanced features greatly increase the percent of an application that can run at vector speeds ("vectorization ratio") for typical application programs.

As a result, the Amdahl vector processors were designed to provide high performance, ease of programming, ease of migration, and high reliability.

- High performance, which results from ultra-high-speed vector and scalar processing and the overlap of vector, scalar, and I/0 processing. The vector processors derive their high performance from very dense packaging of very fast LSI technology. The high-speed vector processing is supported by a fully comprehensive vector architecture integrated with a scalar processor directly compatible with IBM mainframe computers.

- Easy programming environment. Automatic vectorizing through a FORTRAN 77 compiler, a large comprehensive scientific subroutine library, and the availability of a full-screen interactive vectorizer and a full-screen interactive debugger.

- Easy introduction through full IBM compatibility. The vector processor can be attached to any S/370 compatible mainframe running MVS. It attaches to an MVS front-end with a channel-to-channel coupler and shared DASD. Also, since the vector processor is air-cooled, it is physically easy to install.

- The high reliability of the vector processor arises from proven components, the quality control applied to circuitry and packaging, the error-correcting memory, and an error-detection circuit that allows automatic reconfiguration to an alternative memory chip whenever a solid memory failure occurs. Air cooling not only simplifies installation, but reduces operational and maintenance costs.

TECHNOLOGY

The gate-array ECL logic chips contain 400 gates per chip, and some special functional LSIs such as the registers contain 1,300 gates. Signal propagation delay per gate of the LSIs is 350 pico seconds for both types. The high-speed memory LSI contains 4–K bits per module with an access time of 5.5 nanoseconds. They are used where extreme high speed is necessary. Up to 121 LSIs can be mounted on a 29–cm-by–32–cm 14–layered printed circuit board called an MCC (multichip carrier). Logic LSI can be mixed on the same MCC. Thirteen such MCCs are mounted horizontally in a $(30.cm)^3$ cube, called a stack. Forced-air cooling is used throughout the system. With these technologies we can realize a 7.5–nanosecond clock for the vector unit and a 15–nanosecond clock for the scalar unit.

The main storage employs 64–K bit MOS static RAM LSI with 55–nanosecond-chip access time. The maximum size of the main storage unit for models 1200 and 1400 is 256–M bytes. The 500 and 110 have a maximum memory size of 128–M bytes (see Table 7.1).

ARCHITECTURE

This section describes the structure and basic functions of the vector processor. Advanced features will be described separately in a later section. The vector processor shown in Figure 7.1 consists of the scalar unit, the vector unit, and the main storage unit.

Scalar Unit

The functionally independent scalar processor is supported by a large number of general-purpose registers and by a very large buffer storage, or cache, loaded directly from main memory. The peak performance of the scalar processor is in the range of 8–11 megaflops.

The scalar unit fetches and decodes all the instructions. There are 280 instructions, of which 197 are scalar type and 83 are vector type. When an instruction is the scalar type, it is executed in the scalar unit; otherwise it is passed to the vector unit for execution. The scalar unit has 16 general-purpose registers, eight floating-point registers, and 64–K bytes of buffer memory. The floating point registers are 64 bits wide.

Vector Unit

The vector unit of the Model 1200 consists of six functional pipeline units, a 64–K-byte vector register cache, and a 1,024–byte mask register. See Table 7.1 for a description of other models. The functional pipeline units are add-logical pipe, multiply pipe, divide pipe, mask pipe, and two load-store pipes. The first three pipes are for arithmetic operations, any two of which can operate concurrently. All the floating-point arithmetic operations are performed in double-precision (64 bits). The operands and the results for all the vector arithmetic operations are fetched or stored into the vector registers. As shown in Table 7.2, the number of vector registers and their length can be varied to optimize performance dynamically for each loop being executed. Load-store pipes take care of data transfer between main storage and the vector registers; they are both bidirectional.

For model 1200, the throughputs of the add-logical pipe and the multiply pipe are 285 MFLOPS each, whereas that of the divide pipe is 41 MFLOPS. Hence, the 571–MFLOPS maximum throughput is achieved when the add-logical and the multiply pipes are linked together. Each of the two load/store pipes has a data bandwidth of 32 bytes/14 nanoseconds or, equivalently, 2.1 gigabytes/second in either direction. This

Figure 7.1
Vector Processor Model 1200 Block Diagram

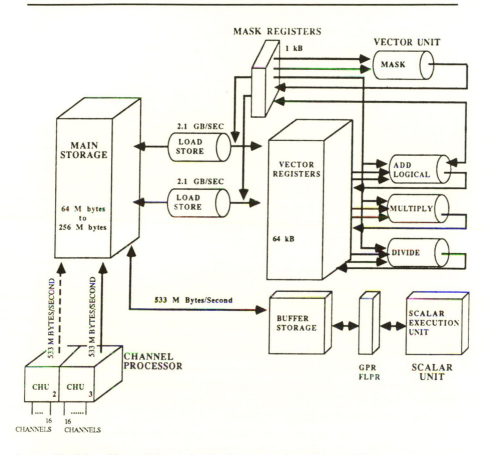

Source: All of the tables and figures in this chapter are from the Amdahl Corporation.

Table 7.1

Hardware Specifications for Amdahl Vector Processor Systems

Feature	VP 500	VP 1100	VP 1200	VP 1400
FCS	3085	3085	3085	2086
Performance: (MFLOPS)				
Peak System	142	285	571	1142
Peak per CPU	142	285	571	1142
Sustainable per CPU	110	220	440	770
Rel. Thghput Capacity	.8	1.2	1.8	2.7
CPU Features:				
Number of CPUs	1	1	1	1
Vector Cycle				
Time (ns)	7.0	7.0	7.0	7.0
Technology (type)	ECL	ECL	ECL	ECL
Technology				
(gates/chip)	400/1300	same	same	same
Data Format Type	IBM	IBM	IBM	IBM
Con. F.P. Pipelines	1	2	2(*)	2(**)
Vector Register:				
Space (KB)	32	32	64	128
Number	8-256	8-256	8-256	8-256
Length	512-16	512-16	1024-32	2048-64
Main Memory:				
Maximum Size (M bytes)	128	128	256	256
# Ports to CPU	1	2	2	1
Cycle Time (ns)	110	110	110	110
Interleaving (ways)	128	128	256	256
I/O System:				
Max Bandwidth				
(M bytes/sec)	96	96	96	96
Max # Channels	32	32	32	32
Channel Speeds				
(M bytes/sec)	3	3	3	3
Max DASD Spindles	>4000	>4000	>4000	>4000
Max M bytes/spindle	1260	1260	1260	1260
Avg Access Time (ms)	23	23	23	23
Max # Tape Drives	?	?	?	?
Auxiliary Memory:				
Option Available (?)	No	No	No	No
Max Size (M bytes)	--	--	--	--
Transfer Rate				
(M bytes/sec)	--	--	--	--

* Each pipeline produces 2 results/7.5 ns cycle
** Each pipeline produces 4 results/7.0 ns cycle

rate matches the maximum throughputs of the arithmetic pipelines. The large capacity of the vector registers can greatly reduce the data traffic to and from the main storage unit. Model 1100 has throughputs for the pipeline units and the total size of the vector and mask registers that are half that of the figures given above.

In order to control conditional vector operations and vector editing functions, bit strings (called mask vectors) are also provided. Mask registers contain one bit for each element in the vector registers, and the mask pipe performs logical operations associated with the mask vectors. (Mask register usage is described later in the discussion of Figure 7.5.)

Main Storage Unit

As mentioned earlier, the maximum capacity of the main storage is 256–M bytes for 1200 and 1400 models with 256–way interleaving, and

Table 7.2
Vector Register Configurations for Amdahl Vector Processors

Number of registers	VP 500	VP 1100	VP 1200	VP 1400
256	16	16	32	64
128	32	32	64	128
64	64	64	428	256
32	128	128	256	512
16	256	256	512	1024
8	512	512	1024	2048

NOTE: Column entries give the maximum number of 8-byte (64-bit) elements a register can hold. For example, the model 500 can have 256 registers each capable of holding 16 elements, or 128 registers capable of holding 32 vector elements, and so on.

128–M bytes with 128–way interleaving of the 500 and 1100 models. Possible vector accesses are contiguous, constant strided, and indirect addressing.

ADVANCED FEATURES IN ARCHITECTURE

This section describes how the advanced features, as listed previously, have been implemented in the vector processor.

Vectorization Level and Program Execution Performance

If VR is the peak vector processing rate for a given application, and SR is the scalar processing rate, then the effective rate EFF is given by Amdahl's law as follows:

$$EFF = \frac{1}{f/VR + (1-f)/SR}$$

where f = fraction of results obtained in the vector unit. Simplifying, we have

$$EFF = \frac{VR.SR}{f.SR + (1-f).VR}$$

Table 7.3
Vector Processing Ratio and Program Execution Performance Values of P for Fraction of Results Obtained in Vector Unit

f (%)	P (times) [a]
0	1.0
10	1.11
20	1.25
30	1.43
40	1.47
50	2.0
60	2.5
70	3.33
80	5.0
90	10.0
99	100.0

[a]Assuming E is infinite.

The performance ratio, P, is the ratio of EFF to SR, that is, the speedup relative to the scalar processing rate,

$$P = \frac{VR}{f.SR + (1-f).VR} = \frac{E}{(1-f)E + f} \tag{1}$$

Where E = VR/SR. Table 7.3 shows the values of P, assuming that E is infinite.

From Table 7.3, we see that if the vector-processing fraction, also known as the vectorization level, is about 50 percent, the performance improvement ratio is two at the highest no matter how large the value of E. This demonstrates that the vectorization level greatly affects program execution performance. To utilize a vector machine to its fullest capacity, it is necessary to develop software that maximizes the vectorization level.

Analysis of Scientific Application Programs

An analysis of the operational characteristics of application programs coded in FORTRAN(1) is shown in Table 7.4.

Table 7.4
Operational Characteristics of Application Programs Coded in FORTRAN

Number	Operation Type	Remarks (Example)
1	Simple DO loop Arithmetic operation	Ai = C * Bi + S
2	Macro operation such as inner product	X = X + Ai*Bi i
3	Indirect accessing and Integer operation	Di = A(Li)
4	Comparative operation or Logical Operation	Li = Ai .GT. Bi
5	Conditional operation	IF (...) Ai = Bi + Ci
6	Index	Ai = 2 * i
7	Gather/scatter	Gathers or scatters data specified by conditions
8	Search	IF (...) GO TO n
9	Others	Recursive operation patterns and scalar processing, etc.

Figure 7.2 shows the percentage of such operational components in the FORTRAN programs analyzed. Figure 7.2 indicates that about half of a typical FORTRAN program performs simple operations such as simple DO loops and arithmetic operations and that the rest of the program performs complicated operations such as those containing control statements (IF statements).

Vectorization Techniques of FORTRAN 77/VP

A vectorizing compiler, FORTRAN 77/VP, has been developed for the vector processor. FORTRAN 77/VP is based on the ANSI 77 standard and accepts all compatible features of ANSI 66 as well. Consequently, it handles all source codes compatible with IBM FORTRAN IV and VS FORTRAN and many additional features. It is equipped with features allowing both global and local analysis and optimization of a program.

In order to obtain high vectorization ratios for a wide range of application programs, the FORTRAN 77/VP compiler vectorizes not only simple DO loops, but nested DO loops and such macro operations as the

Figure 7.2
The Percentage of Operational Components in FORTRAN

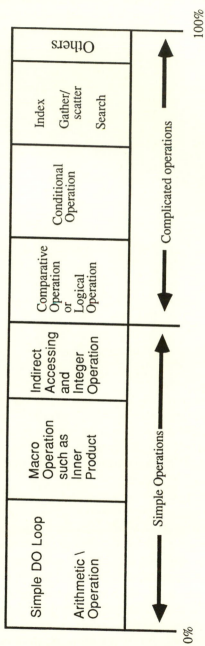

Table 7.5
Vectorization Techniques Classified by Program Structure

Vectorization Technique	Program Structure	Operation Type (from Table 7.4)
Vectorization of basic operation	Simple DO loop	1,3,6
Vectorization of simple variable	Simple DO loop	1,2
Vectorization of macro operation	Complicated control flow	2,7,8
Vectorization of IF statement	Complicated control flow	4,5
Vectorization of nested DO loops	Nested DO loop	--

inner product efficiently. It also detects and separates recursive operations.

As shown in the previous section, it is obviously impossible to maximize the power of a vector machine by simply performing vector processing only for simple operations. That is, several vectorization functions must be thoroughly implemented in the compiler to increase the vectorization level. The following items outline the vectorization functions of the FORTRAN 77/VP compiler.

Operation Types and Vectorization Techniques. Vectorization techniques are classified as shown in Table 7.5 according to program structures and the types of data that appear in DO loops in FORTRAN programs (operation types).

Vectorization of Basic Operations. In order to get high vectorization levels, it is necessary to minimize the number of scalar operations.

Whether or not a DO loop can be vectorized depends on the data dependence within the loop.[2] If there exist recursive data dependencies in a loop, then the loop cannot be vectorized. Using the SCC (strongly connected components) detection algorithm,[3] the compiler distributes the operations in a DO loop to vectorizable and nonvectorizable operations. We call this technique "partial vectorization." Figure 7.3 illustrates how this operation is handled. Note the expressions for vectors A and X are moved out of the loop. Evaluation of these two vectors is performed with vector instructions. However, we cannot move the evaluation of vector D ahead of the loop since values of D depend on com-

Figure 7.3
Partial Vectorization

```
Source Program

DO 20    I=2,N
    X(I)   =B(1)+C(I)
    B(I)   =X(I)/E(I)
    E(I+1) =B(I)+C(I)
    D(I)   = (E)I)+1.0)**2
    A(I+1) =C(I)-1.0

20 CONTINUE
```

$$\Downarrow$$

```
                              Vectorization

V    A_{i+1}=C_1-1.0          ,i=2,3,...,N
V    X_i   =B_i+C_i           ,i=2,3,...,N
S    DO 20   I=2,N
S      B(i)=X(I)/E(I)
S      E(1+I)=B(I)+C(I)
S    20 CONTINUE

V    D_i=(E_i+1.))**2         ,i=2,3,...,N
```

---

puted values of vector E. Computation of vector D is therefore performed after the loop.

Vectorization of Macro Operations. Macro operations are a series of operations executed as one procedure in a DO loop. The calculation of a vector inner product is an example of a macro operation.

DO 40 I = 1,N

X = X + A(I)*B(I)

40 CONTINUE

Other examples of macro operations are the computation of vector summation and searching for maximum-minimum values. The FORTRAN 77/VP compiler detects cyclically defined simple variables to identify the appropriate macro operation.

Vectorization of IF Statement. In scientific programs, IF statements frequently appear in a DO loop. It is therefore necessary to vectorize IF

Table 7.6
Frequency Percentage of IF Statement Forms and Action

If statement form	Percentage	
IF (invariant expression in the loop)	15	Unchanged
IF (expression explicitly containing DO control variable	19	Conversion of IF statement by applying the range of DO control variable to executable statements
IF (general variable expression)	66	Vectorization of general IF statement

statements effectively in order to achieve high performance rates. The following discussion deals with IF statements containing a conditional expression whose value is not changed recursively by the execution of a DO loop. The percentage of each IF statement form and the corresponding action taken by the compiler are listed in Table 7.6.

In Figure 7.4 we show how FORTRAN 77/VP vectorization nested IF statements. :M indicates the mask function that enables or disables the operation at the left according to each value of M (true or false).

Now we will take a look at how instructions are generated for this kind of conditional operation. In the VP system, the gather-scatter or compress-expand method and the list vector method are also available for the handling of conditional operations, in addition to the mask operation given above.

Therefore conditional operations should be converted to hardware instructions not necessarily via mask operations but more flexibly.

Figure 7.5 shows how the conditional expression $A_i + B_i: M_i$ ($i = 1,2, \ldots 8$) is processed in these three methods. The following describes the figure:

- *Masked Operation Method*: The masked operations available with the hardware are used. Although only the true elements of the masked part are processed, the execution time is the time to process all vector elements.

- *Gather-Scatter Method*: The elements to be processed are gathered prior to execution. The result is scattered into the original vector. In this case, the overhead of gather-scatter load-store operations is incurred.

- *List Vector Method*: An index list of the elements to be processed is created. The elements to be processed are retrieved into vector registers according to the index list.

Figure 7.4
Vectorization of IF Statements

```
                Source Program

        DP 20    I=1,N
          A(I)     =A(1)+1.0
          IF       (A(I).GT.B(I))    GO TO 5
          B(I)     =C(I)*X
          IF       (B(I).GT.Y)       GO TO 6
     5    C(I)      =D(I)+B(I)
     6    D(I)      =E(I)/Z

    20 CONTINUE
```

⇓

```
                                        Vectorization

        Ai=Ai+1.0                       ,i=1,2,...,N
        M1i≠Ai.GT.Bi                    ,i=1,2,...,N
        M2i=.NOT.M1i                    ,i=1,2,...,N
        Bi=Ci*X                :M2i     ,i=1,2,...,N
        M3i=Bi.LE.Y            :M2i     ,i=1,2,...,N
        M4i=M3i.OR.M1i                  ,i=1,2,...,N
        Ci=DiBi                         ,i=1,2,...,N
        Di=Ei/Z                         ,i=1,2,...,N
```

The approximate execution time is calculated for each method and the method that takes least execution time is selected. The execution time for the masked operation method (T_m), the gather-scatter method (T_g), and the list vector method (T_l) are given by these formulas:

$$T_m = T_{mls} + T_{ex} \tag{2}$$

$$T_1 = T_{lls} + t*T_{ex} + T_{laux} + T_{vl} \tag{3}$$

$$T_1 = T_{gls} + t*T_{ex} + T_{gaux} + T_{vl} \tag{4}$$

Where,

T_{mls}: Execution time of load-store for masked operation method

T_{lls}: Execution time of load-store for list vector method

T_{gls}: Execution time of load-store for gather-scatter method

T_{ex}: Execution time of operation

T_{laux}: Execution time for generating list vector

T_{gaux}: Execution for gathering and scattering

Figure 7.5
Outline of Three Conditional Operation Methods

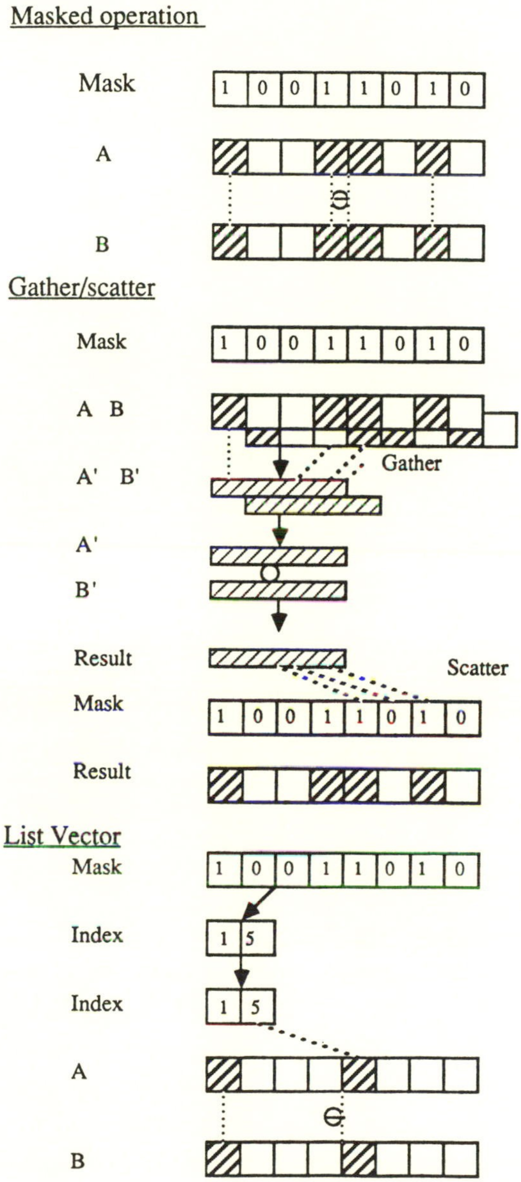

Masked operation

Mask   1   0   0   1   1   0   1   0

A

B

Gather/scatter

Mask   1   0   0   1   1   0   1   0

A   B

A'   B'   Gather

A'

B'

Result   Scatter

Mask   1   0   0   1   1   0   1   0

Result

List Vector

Mask   1   0   0   1   1   0   1   0

Index   1   5

Index   1   5

A

B

Figure 7.6.
The Rule of Selecting the Optimum Method

LOAD STORE density

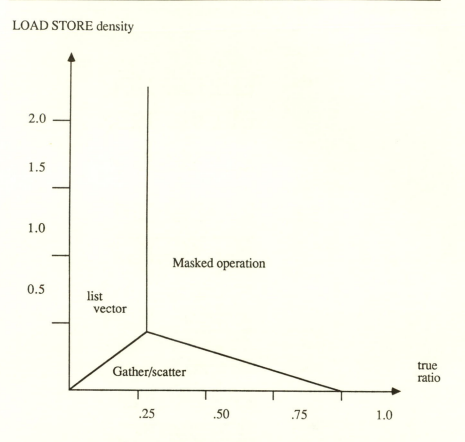

T_{v1}: Execution time for changing vector length

t: True ratio of logical expression

Figure 7.6 shows the rule of selecting the optimum method. [The horizontal axis shows true ratio (t), and the vertical axis shows density of load-store (number of load-store operations divided by number of operations)].

Vectorization of Nested DO Loops. In scientific application programs, nested loops are commonly encountered. To increase the vector-processing ratio, it is necessary to vectorize the noninnermost loops; in addition, it is desirable that the innermost loop be vectorized, not with the innermost index only, but with optimal indices. To vectorize noninnermost loops, loop distribution is necessary (see Figure 7.7).

Figure 7.7
Loop Distribution

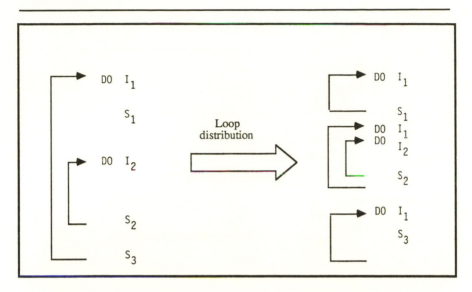

Loop distribution can be performed by checking data dependence in the same manner as for the vectorization of basic operation. In Figure 7.7, loop distribution is possible unless the data dependence with respect to index I_1 forms a loop among operations S_1, S_2, and S_3.

A loop where no executable statement appears between DO statements, such as the loop in S_3 after distribution, is called a *tight loop*. When vectorizing a tight loop, several methods are used to increase the efficiency of vector operation—for example, not only the innermost index but also an outer index is selected for vectorization or multiple loops are grouped. Vectorization of grouped loops is called "volume vectorization." In some cases, nested loops can be vectorized via linear mapping of multiple indices even when they cannot be vectorized with the innermost index.[4] Figure 7.8 shows an example of the vectorization of typical nested loops.

In Figure 7.8, equation (5), a vector operation can be performed for the entire vector since the range of the DO index equals the size of the array (volume vectorization). In equation (6), vectorization, if performed after loop distribution, will produce an inner product; the best DO index (which allows vector data to be contiguously accessed) is selected for efficient vectorization (index permutation). Note that loops with indices J and K have been switched to allow column accesses in the innermost loop. Because of the FORTRAN conventions regarding storage of matrices, column accesses are much faster than row accesses.

Figure 7.8
Nested Loops

```
        Source Program
        DO 90 I=1,N
           DO 90 J=1,N
              A(I,J)=0
              DO 90 K=1,N
                 A(I,K)=A(I,J)+B(I,K)*C(K,J)
 90 CONTINUE
```

$$\Downarrow \quad \text{Loop distribution}$$

```
        DO 90 J=1,N
           DO 90 I=1,N
              A(I,J)=0
        90 CONTINUE
        DO 91 I=1,N
           DO 91 J=1,N
              DO 91 K=a,N
                 A(I,J)=A(I,J)+B(I,K)*C(K,J)
 91 CONTINUE
```

$$\Downarrow \quad \begin{array}{l}\text{Vectorization of a tight loop}\\ \text{Index Permutation}\end{array}$$

```
        A_i = 0                         i'=1,1,...,N*N    (5)
        DO 91 K=1,N
           90 91 J=1,N
              A_{i,J}=A_{i,J}+B_{i,K}*C_{K,J}   i=1,2,...,N   (6)
 91 CONTINUE
```

Pipeline Parallelization. The vector processor system has three arithmetic pipelines, two memory access pipelines, and one mask pipeline, the mask pipeline and the scalar execution unit can operate concurrently. Pipeline parallelization uses this feature and produces an object code that allows vector operations to be executed concurrently during instruction scheduling. It analyses the data dependence of FORTRAN statements and resequences operations so that vector and scalar operations can be overlapped while maintaining any existing data dependencies between vector and scalar data.

The way concurrent vector operation increases overall efficiency is shown in Figure 7.9.

Dynamically Reconfigurable Vector Registers. One unique feature of the vector processor is the dynamically reconfigurable vector registers. The analysis of the 1,000 FORTRAN programs indicated that the requirements for the length and the number of vector registers vary from one program to another, or even with a program. To make the best use of the total capacity of 64–K bytes for the 1200, for example, the vector registers may be concatinated to take the following configurations: 32 (length) × 256 (total number), 64 × 128, 128 × 64, ... , 1,024 × 8 (see Table 7.2). The length of vector registers is specified by a special register and it is set by an assembler instruction generated by the compiler. The FORTRAN compiler will attempt to assign optimal register lengths for each loop.

In order to best utilize the dynamically reconfigurable vector registers, the compiler must know the frequently used vector length for each program, or even within one program the vector length may have to be adjusted. When the vector length is set too short, load-store instructions will have to be issued more frequently, whereas if it is set unnecessarily long, the number of available vectors will decrease, resulting in frequent load-store operations again. In general, the compiler puts a higher priority on the number of vectors rather than on the length in searching for the optimal register configuration.

PROGRAMMING CONSIDERATIONS

The FORTRAN 77/P system provides the user with these features to facilitate programming:

- VSOURCE listing. The vectorization indicator is printed on the VSOURCE listing to inform the user of vectorized statements. The character V preceding a statement indicates that this statement will be executed in the vector unit. An example of a VSOURCE list is shown in Figure 7.10.

- Compiler output messages. The compiler outputs vectorizing messages indicating how statements have been vectorized and why certain statements in

Figure 7.9
Effects of Pipeline Parallelization

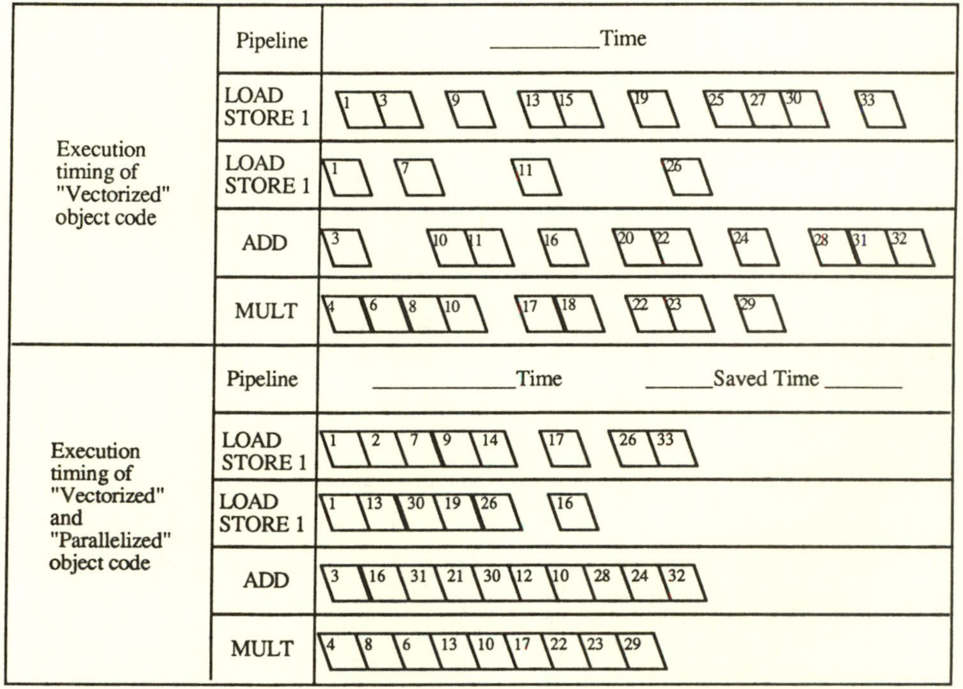

	Pipeline			Time							
Execution timing of "Vectorized" object code	LOAD STORE 1	1	3	9	13	15	19	25	27	30	33
	LOAD STORE 1	1	7		11		26				
	ADD	3	10	11	16	20	22	24	28	31	32
	MULT	4	6	8	10	17	18	22	23	29	

	Pipeline			Time				Saved Time			
Execution timing of "Vectorized" and "Parallelized" object code	LOAD STORE 1	1	2	7	9	14	17	26	33		
	LOAD STORE 1	1	13	30	19	26	16				
	ADD	3	16	31	21	30	12	10	28	24	32
	MULT	4	8	6	13	10	17	22	23	29	

A parallelogram indicates execution timing of a vector instruction.

Figure 7.10
VSOURCE List

```
FORTRAN 77/VP VIOLIOC MAXIMM     DATE 82.10.23     TIME 16:27.15

   LNO      V                 SOURCE
00000200              SUBROUTINE MAXIMM(XTAB,XMAX,N,INCX,ISAMAX)
00000210      C
00000300      C  *******FIND SMALLEST INDEX OF REAL VECTOR COMPONENT
00000400      C           OF MAXIMUM MAGNITURE              *********
00000410      C
00000500              REAL XTAB(N), XMAX
00000510      C
00000520      C
00000600              ISAMAX = 1
00000700              IF ( N .LE. 1 )  GO TO 20
00000800                  MAX = ABS( XTAB(1)    0
00000900                  IX = INCX + 1
00001000  V               DO 10 I = 2,N
00001100  V                   IF ( ABS( XTAB(1X) ) .LE. XMAX ) GO TO 10
00001200  V                       ISAMAX = I
00001300  V                       XMAX = ABS ( XTAB(1X) )
00001400  V       10              IX = IX + INCX
00001500         20     CONTINUE
00001600                RETURN
0000170                 END
```

the source code cannot be vectorized. The compiler will also issue tuning messages indicating how statements can be restructured to improve source code efficiency.

INTERACTIVE VECTORIZER

Although the FORTRAN 77/VP compiler usually does an excellent job of vectorizing a program with no help from the programmer, there are cases where vectorization is inhibited because the compiler lacks information. This information can include the iteration count for a DO loop, the value or sign of a DO loop increment, or relationships between array subscripts. The FORTRAN 77/VP compiler will accept directives that provide the necessary information, but the user is often unsure of what information is most effective and where to put it. The compiler directives are known as vector optimization control lines (VOCL) and appear to other FORTRAN compilers (including FORTRAN 77) as comments. The interactive vectorizer tells the user what directives are required and where to put them. It also performs an analysis that shows how well program units have been vectorized and what the effect of the VOCL will be.

The interactive vectorizer operates in three modes: a dynamic mode, a static mode, and a unit mode. In the dynamic mode information provided by FORTUNE is read by the vectorizer. This provides it with the precise values for DO loop iteration counts, the number of branches through IF statements, and the number of executions of each program unit. The dynamic analysis is the most precise available provided that the FORTRAN run was performed with typical data. The static analysis is used when a FORTUNE run is not available; for example, when an incomplete program is being analyzed. In this case, the vectorizer assumes values for DO loop iteration counts and other unknown quantities. The unit mode is a corner variation of the static mode.

For each program unit being analyzed, the vectorizer computes the cost of each statement before and after vectorization. This information is used to build a DO loop list that shows the cost of the vectorized and unvectorized loop. With the loop list displayed on the screen, the programmer can split the screen and display the output from the vectorizing compiler (FORTRAN 77/VP) in the bottom half of the screen. Selection of a specific loop from the loop list will produce the vectorization messages for the loop. In the case of messages for loops that do not vectorize very well, the user can ask for more information. If the compiler lacks information that would enable it to vectorize the loop, the vectorizer tells the user what kind of VOCL is required and where to put it in the source. The user can edit this directly into the source, without leaving the interactive vectorizer. Once inserted, the effect of the VOCL can be determined, at both the statement and program unit level.

The interactive vectorizer has a batch counterpart, which performs all the analysis performed by the interactive vectorizer, but the user cannot have the dialogue that follows this in the interactive mode. Figure 7.11 gives a sample of interactive vectorization.

FORTUNE

FORTUNE is a performance measurement and turning tool that actually runs the program to determine how many times each statement was executed for a specific set of data. It consists of two parts—an editor and an analyzer. The editor inserts counters into the program. The program is then compiled by FORTRAN 77 and executed. During execution the counters are incremented as statements are executed. The analyser uses the values of the counters to estimate the relative costs of each statement in the program units. The summation of the appropriate counters provides the relative cost of each program unit and also the number of times each was executed. The counters may be accumulated in a user-supplied data set so that the performance of the program over a variety of sets of data may be determined. Not only does FORTUNE enable the programmer to determine where the most time-consuming

Figure 7.11
Sample of Interactive Vectorization

```
18900                   DO 100 J=1,N

19000 V                    DO 100 I=2,N-1

19100 V                        DKX(I,J)=DKY(I,J)-X*(0.5*(DEX(I+1,J)-

19200          *                                     DEX(I-1,J)) +

19300          *                    DEY(I,J)-DEY(I-1,J))/V(I,J)

19400 V                 100 CONTINUE
```

```
   - LNO 19000 - 194000 VECTORIZED BY INDEX I.

   - IF YOU REWRITE THIS LOOP USING MASKED VECTOR,
     THESE TIGHT LOOPS WILL BE VECTORIZED TOTALLY.
```

```
20100 V                 DO 200 J=1,N

20200 V                    DO 200 I=1,N

20300 V                    IF (I.EQ.1 .OR. I.EQ.N)  GO TO 200

20400 V                        DKX(I,J)=DKY(I,J)-X*(0.5*(DEX(I+1,J)-

20500          *                                     DEX(I-1,J)) +

20600          *                    DEY(I,J)-DEY(I-1,J))/V(I,J)

20700 V      200 CONTINUE
```

parts of the program are, it also indicates which parts of the code are infrequently or never used. The latter may enable the programmer to determine if such parts of the program have been adequately tested.

DOCK/FORT

DOCK/FORT 77 is a full-screen interactive debugger that provides conversational debugging of FORTRAN programs. The user can watch the program execute and can control the speed at which each program unit executes. The screen is managed in such a way to enable the user to display the source of the program unit being executed and at the same time display the results from the program and messages from DOCK/FORT 77 alongside it. Traps may be set to stop execution of the program when a user-specified condition, such as a variable changing its value, occurs. If this tool is being run on a color terminal, the last statement executed and the next to be executed are highlighted in a different color to the rest of the code. The user may change the values of variables, change the execution sequence "on-the-fly," may add temporary FORTRAN statements, and ask for the values of variables in the program as it executes. DOCK/FORT 77 provides the user with a permanent copy of all the changes and output produced during the session if it is required. DOCK/FORT 77 compiles the user's program with FORTRAN 77 and the compilation step may be executed as a batch job if required.

ACKNOWLEDGEMENT

This chapter is based, in part, on two publications by the staff at Fujitsu Ltd.: (1) Kenichi Miura and Keiichiro Uchida, "FACOM Vector Processor VP–100/VP–200," in *Proceedings of the NATO Advanced Research Workshop on High Speed Computing* (1983), and (2) S. Kamiya, F. Isobe, H. Takashima, and Y. Tanakura, "Practical Vectorization Techniques for the FACOM VP," in *Proceedings of IFIP Congress*, Paris (1983).

NOTES

1. S. Kamiya et al., "Practical Vectorization Techniques for the FACOM VP," IFIP Congress, Paris, September, 1983.
2. D.J. Kuck, Y. Muraoka, and S.C. Chen, "On the Number of Operations Simultaneously Executable in FORTRAN-like Programs and Their Resulting Speed Up," *IEEE Transactions on Computers*, C–21; D.J. Kuck, "A Survey of Parallel Machine Organization and Programming," *Computing Surveys* 9 (1977): 29–59; and D.A. Pauda, D.J. Kuck, and D.H. Lawrie, "High-Speed Multiprocessors and Compilation Techniques," *IEEE Transactions on Computers*, C–29 (1980): 763–776.
3. A.V. Aho, J.E. Hopcroft, and J. Ullman, *The Design and Analysis of Computer Algorithms* (Reading, MA: Addison-Wesley, 1972), 189–195.
4. L. Lamport, "The Parallel Execution of Do Loops," *Communications of ASM* 17 (1974): 83–93.

8

WHY PARALLEL PROCESSING IS INEVITABLE

W. DANIEL HILLIS

All technologies improve through successive waves of new developments, each of which supplants the previous wave and carries forward to new levels of price performance. High-end scientific computers are a prime example: Vector processors completely replaced serial machines as they ran out of growth potential in the late 1970s. Now, as vector processors run out of growth room, parallel processors are poised to make the next leap forward. Thinking Machines Corporation is leading this new wave with its massively parallel Connection Machine system.

There is nothing inevitable or permanent about any design in a fast-moving technology like computer systems. What we use today is what is most cost-effective today. It replaced what we used to use yesterday and will in turn be replaced by what we will use tomorrow. An example from outside our industry illustrates this point. Figure 8.1 is a chart of particle-accelerator performance. The vertical axis is the maximum amount of collision energy available. What it shows is an exponential increase in the amount of power available. But a closer look at the details reveals that there were different types of machines used over time. Different technologies came in one at a time. As one technology matured, it began to reach a limit in its ability to give more energy and it was replaced by another technology. The result is the "head and shoulders" phenomenon characteristic of any rapidly developing technology. This is exactly what is happening right now in the high end of the scientific computer industry. As shown in Figure 8.2, the first stages of growth came from speed ups in serial processors. They are analogous to the Van de Graaff generators in Figure 8.1. After a while, of course, the Van de Graaff generators were replaced by cyclotrons. In a similar way, serial scientific computers were replaced by vector processors. These vector

Figure 8.1
Particle Accelerator Performance

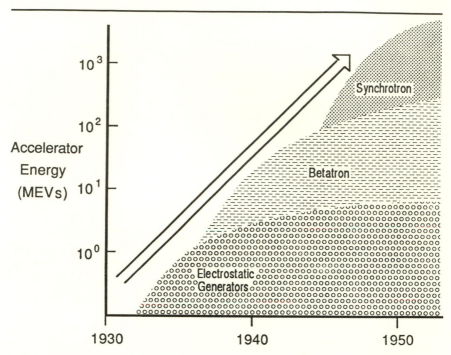

Source: Compiled by the author. For more detail on this phenomenon, see Gerald Holton,
   Thematic Origins in Scientific Thought (Cambridge, MA: Harvard University Press,
   1973), p. 416.

processors are now giving us a level of performance that simply cannot
be achieved on a serial machine.

We now have parallel processors that are just beginning to cross over
in one of those leaps and a few years from now we will have performance
that is hundreds of times more than is possible with vector processors.

It is worth reviewing the reasons, the fundamentally correct reasons,
for going to parallel processing. Computers today are built essentially
on the von Neumann architecture—an essentially two-part architecture
with memory on one side and processing on the other side. A modern
supercomputer has about a square meter of silicon. Almost all of it is
used for memory processing. Figure 8.3 illustrates the silicon utilization
in a contemporary scientific supercomputer. In a vector processor there
is a high utilization of the switching circuits in the processor area, but
most of the memory is idle most of the time. It is fundamentally an
inefficient design. That is something everyone realizes, which is why

Figure 8.2
High-End Computer Performance

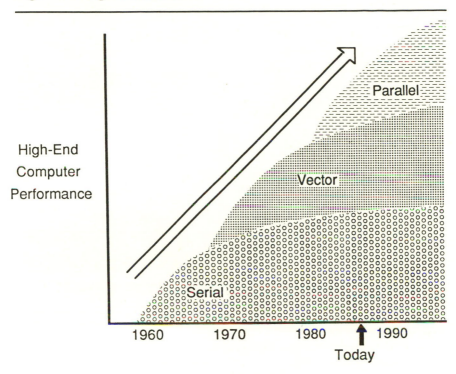

Source: Compiled by the author.

the technology went to vector processing in the first place. It raises the overall utilization from a 0.10 percent or so to about 1 percent. But there is still a long way to go.

Clearly the answer will lie in somehow spreading the computation out over the silicon. The flavor of the answer is going to be in mixing the processing and the memory. Another way of doing this is "parallel processing"—having smaller units of processing power working on the problems simultaneously. This is going to be the solution because the speed-of-light limitation forces it. Figure 8.4 illustrates this alternative way of structuring the silicon.

Fortunately, we can structure our problems that way because the world works that way. The world is much more like Figure 8.4 than it is like Figure 8.3. A specific example of parallel processing that does not involve four conventional processors, or even 40, is multiprocessing. The significant advances we need require thinking about the problems in a completely different way. Machine visions provide a clear example.

Figure 8.3
Silicon Utilization in a Contemporary Supercomputer

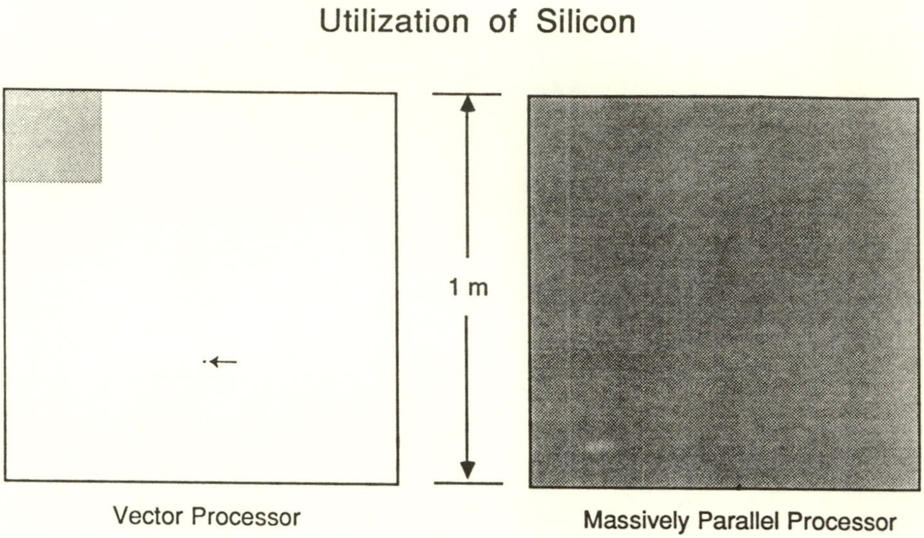

Utilization of Silicon

Vector Processor   Massively Parallel Processor

The human retina does a recognition problem, or an inspection problem, by having a processing element for each point in the retina. The analogous solution on a parallel processor is to use a processor for each element, or pixel, in the image. A 1,000 × 1,000 pixel image requires 1 million processors. The kinds of processors that are being developed in the Strategic Computing Program of the Defense Advanced Research Program Agency and constructed at Thinking Machines really do provide the user with that number of processing elements.

However, unlike the vision problem in which a million processors are connected in a two-dimensional grid resulting in a factor of a million in parallelism, most computation-intensive problems aren't arranged exactly the way the processors are connected. For example, to stimulate the behavior of a VLSI in an integrated circuit with several hundred thousand transistors, the natural connectivity of the problem depends on the connectivity of the circuit being simulated. If one transistor were connected to another transistor, it would be optimal for the processor that simulates the transistor to be connected to the processor that simulates the other one. One processing element is used for each simulated element, so if the circuit has 100,000 transistors, it will be simulated using 100,000 processors. For this example the Connection Machine would actually dynamically reconfigure itself in a few milliseconds so

Figure 8.4
Parallel Processing—An Alternative Way of Structuring Silicon

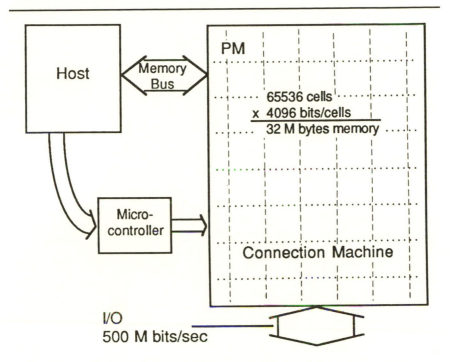

Source: W. Daniel Hillis, *The Connection Machine* (Cambridge, MA: MIT Press, 1985), p. 21. Used by permission of the author.

that the processors are wired in the same pattern as the circuit diagram. Then each processor just has the problem of simulating a single transistor and the whole problem can proceed 100,000 times faster.

The same technique can be applied, for instance, to language understanding data structures to the higher-level semantic structures of vision to signal processing. Typically for each stage of computation in this type of problem, the natural pattern rearranges itself so that, for example, the vision problem may start out with a gridlike computation followed by a treelike computation in the next stage. As the stages of computation progress to higher features, a graphlike structure would emerge, and the Connection Machine would take just a few milliseconds to reconfigure itself, achieving peak parallelism at each stage of the problem.

It is important also to look at the transition from the software perspective. It is legitimate for people to feel that the leap in parallel processing is just in the process of happening. Vector processing still provides the most reliable computrons per dollar right now. The experimental systems that leap over vector processing are just beginning

Figure 8.5
Representation of a Single Virtual Copy Network

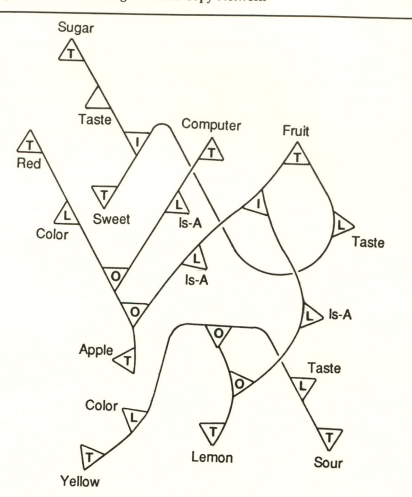

Source: W. Daniel Hillis, *The Connection Machine* (Cambridge, MA: MIT Press, 1985), p. 19. Used by permission of the author.

to be demonstrated. Right now they are approximately in the same area, but the leap in software occurred five years ago. The tools that started to become available then have allowed programmers to write much more complex systems successfully.

Figure 8.5 illustrates a common-sense reasoning program that must make deductions based on knowledge of hundreds of thousands of facts. This program requires approximately a million lines of code. There is no way that such a program could be successfully implemented in FOR-

TRAN, but these kinds of programs have been successfully written for years now. Essentially, the tools have gotten much better. Getting better productivity in large-scale programming projects requires a changeover to these new kinds of tools.

The long-term risk is the same in both hardware and software. In an environment of staged advances, it is possible to become very competent at an obsolete technology. The chance of this happening is greatest at a time, like today, when a transition to a new stage is just occurring.

PART III

THE ROLE AND SCOPE OF FEDERAL AGENCIES IN SUPERCOMPUTER DEVELOPMENT

INTRODUCTION

JAMES DECKER

The federal government has been the primary customer for super-computers in the past. In fact, most supercomputers were developed to meet the requirements of national security programs. Such government-funded laboratories as the Lawrence Livermore National Laboratory and the Los Alamos National Laboratory acquired the first or one of the first of each new generation of supercomputers. These laboratories invested considerable resources in developing the software required to make us-able supercomputer systems. In addition, a substantial effort was ex-pended to develop the computational techniques necessary to model large, complex scientific and engineering problems. The combined tal-ents of the supercomputer vendors and the government laboratories produced unchallenged world leadership in the development, manu-facture, and application of supercomputers.

The government's requirements for supercomputers over the years have led to the development of increasingly more powerful hardware, improved software, and improved mathematical techniques. Conse-quently, applications of supercomputers have increased from just those associated with national defense to many engineering applications in scientific research. Computational science has become a key element of scientific research along with experiments and analytical theory. Because of the growing importance of computational science and engineering, supercomputers are sometimes called the "machine tools" of the modern high technology world.

In the early 1980s, two issues related to the development and use of supercomputers in this country were raised in many different forums. One issue was the lack of access to supercomputers for all but a few scientists (primarily in government laboratories) in the country. The primary concern with the lack of access was that U.S. scientists were not able to use an important research tool, the supercomputer, and would soon not be competitive with their foreign counterparts. In fact, many stories circulated about scientists from U.S. universities going to foreign laboratories and universities to obtain the supercomputer time necessary to do their calculations.

The second concern was the rate of increase of performance for new generations of supercomputers produced by U.S. vendors. Many large, complex problems encountered in government research and development programs require the use of supercomputers far more powerful than those available today. In the early 1980s, U.S. vendors did not appear to be moving rapidly enough to develop new machines to meet the needs of these programs. In addition, three Japanese companies had announced their intention to build supercomputers with performance objectives that might equal or surpass those produced by U.S. vendors. Many groups were concerned that because of the importance of supercomputers to both military and economic security, the United States should not lose its leadership position in this technology.

These concerns about the future development and application of supercomputer technology in the United States led to a number of federal studies, congressional hearings, and symposia. Among the most important of these initiatives were

- the Lax Committee and its report
- hearings held by the House Science and Technology Committee (June and October, 1985)
- formation of the Federal Coordinating Council on Science, Engineering, and Technology (FCCSET)

Most of these studies reached the same general conclusion: The United States was in danger of losing its leadership position in supercomputer technology. The lack of access to supercomputers for many scientists presented a serious problem for the U.S. scientific enterprise.

The FCCSET committee reports and present activities are illustrative. FCCSET is a high-level interagency committee organization by the Office of Science and Technology Policy. In the spring of 1982, FCCSET established a supercomputer panel to review actions the federal government should take to maintain U.S. leadership in the development and applications of supercomputers. The panel in turn formed three working groups to address these issues:

- supercomputer access
- government actions to maintain U.S. leadership in supercomputer development
- interagency coordination of research in very-high-performance computing

The first two groups were asked to and did submit reports by fall 1983. Among the most important recommendations was that the federal government should fund a limited number of university-based super-computer centers and link them through wide-based communications networks. This approach would address the access problem, but take into account the great expense of the computers, their enormous capacity, and the specialized expertise required for their operation. The Department of Energy Center at Florida State University and the four National Science Foundation Centers are a direct outgrowth of this rec-ommendation. The DOE Magnetic Fusion Computer Network, opera-tional since the mid 1970s, had demonstrated the viability of the network approach.

In addition to the recommendations that address access, the FCCSET committee recommended other policies that have been adopted and serve as basic guidelines for federal support of supercomputer computer technology:

- The government should continue to be a friendly buyer of supercomputers from U.S. vendors; that is, the government should continue to commit to buying early models of new generations of machines, even though they may come nearly devoid of software. (The government should continue to develop software as necessary.)

- The government should increase its support for long-range research in new computer architectures and applied mathematics with the objective of signif-icantly increased computing performance.

The FCCSET committee continues to make policy recommendations and coordinate government activities in the supercomputer area. During 1982–1983 a special subcommittee was established to improve the co-ordination of networking supercomputers among the federal agencies, and a memorandum of understanding was proposed to facilitate sharing of supercomputer resources between agencies. Progress in developing future supercomputers by U.S. vendors as well as foreign companies has been monitored; plans for training new supercomputer center per-sonnel as well as new users have been developed and training courses have been initiated. In addition, the subcommittee that coordinates long-range research in high performance computing has recently produced a report that contains a review of all federal programs in this area and contains new policy recommendations as well.

Accordingly, as we enter the second half of the decade, the U.S. government has resumed its historic, activist role in the development of advanced, large-scale computational systems. Such involvement is

once again regarded as a key to maintaining U.S. scientific and technological preeminence.

The chapters in Part III review activities by several agencies that fund
or utilize supercomputers. They are representative of federal activities
in the supercomputer field.

9

THE ROLE OF SUPERCOMPUTERS IN ENERGY RESEARCH PROGRAMS

JOHN KILLEEN

During fiscal year 1984, the Department of Energy created the Energy Sciences Advanced Computation activity and established, as its major program, a supercomputer access program. This program was initiated as the result of various panels that had investigated the availability of modern supercomputer resources to the scientific research community within the United States and to the DOE research community in particular. It was found that the current availability of modern supercomputer resources within the United States fell far short of the amount of these resources needed by the research community, and it was also found that modern supercomputers themselves do not have sufficient capability to address many of the computational needs of this community. During fiscal year 1984 a requirement analysis was conducted throughout the research community funded by the Office of Energy Research, and this analysis verified that several Class VI computer systems would be needed to begin satisfying this suppressed demand.[1]

The disciplines with supercomputering needs include high-energy physics, nuclear physics, chemical and materials sciences, engineering and applied mathematical sciences, geological and meteorological sciences, and biological and related sciences. Although extensive computing requirements in these fields have already been identified, new problem areas are continually being uncovered and the magnitude of the latest demand for supercomputing in Office of Energy Research programs is just beginning to be understood.

The purpose of the Energy Sciences Advanced Computation Supercomputer Access Program is to provide nationwide high-speed-network access to modern centralized facilities within the constraints of budgetary resources. In order to begin addressing this access problem as quickly and as economically as possible, the Office of Energy Research (ER) decided to utilize the existing National Magnetic Fusion Energy Computer Center (NMFECC) and its installed high-speed satellite network across all ER programs. Because the NMFECC satellite network was already accessible at many DOE laboratories and universities and because this network provides gateways to other networks, such as ARPANET and TYMNET, many researchers were able to gain access to the NMFECC facilities with very little lead time and minimal additional cost.

The Office of Energy Research is funding the Cray XMP–2 computer system installed at the NMFECC in November, 1984, to further expand the availability of supercomputer resources to the nonfusion ER programs. This system addresses the near-term capability and capacity needs. The Office of Energy Research is requesting funds for a more advanced Class VII system in fiscal year 1988 in order to provide the capabilities required. The Class VII system will be acquired through a competitive procurement at a time when U.S. vendors are expected to market at least three systems of this capability.

THE NEED FOR MORE POWERFUL COMPUTERS

Scientists who use supercomputers have in the past constrained their numerical simulations to an average execution time of about ten hours. This constraint reflects the scientist's need to make daily progress. Thus the amount of complexity incorporated in models is scaled to the computer's ability to produce results in about a ten-hour execution time. The capability of a supercomputer dictates the amount of complexity that can be treated. Because of this limitation, scientists engaged in large-scale numerical simulation have continually sought bigger and faster computers. Today scientists engaged in energy research need supercomputers that are up to 200 times faster than state-of-the-art equipment.

In order to understand the requirements for more powerful computers, we must explore the generic reasons for having increased computational speed and storage.

Dimensionality: The real world exists in three space dimensions plus time. If computational models reflected the real world exactly and completely, they would treat all four of these dimensions and other parameters that are equivalent to additional dimensions. With current computers, it is possible to treat two space dimensions and time for some problem types, three space dimensions and time for others, and three space dimensions and time for a very limited set of problems. Speed increases of about a factor of 200 in this decade are needed to allow researchers to solve urgent multidimensional problems that are now intractable.

Resolution: Every region of space contains infinitely many points. Thus the first step in modeling any natural phenomenon is to approximate the space with a finite set of zones, each of which requires a number of calculations. Increasing the number of zones means we can determine more completely and accurately what is happening in any environment, but the computational time grows very rapidly. For example, in a two-dimensional time-dependent model, the running time grows in proportion to the third power of the increase in resolution; increasing the number of zones by just a factor of two would increase the time to complete the problem by a factor of eight. Many complex problems now run up to 100 hours, so it is clear that resolution increases of even relatively small factors can overwhelm the capabilities of current supercomputers.

Physics: All computational models dealing with the frontiers of science and technology make simplifying assumptions about the laws of physics in order to keep the calculations from running too long. In some models, including just one additional physical effect can increase running time by a factor of ten. Faster supercomputers with much larger memories will permit researchers to solve problems that cannot now be economically solved.

Combination effects: Although dimensionality, resolution, and physics each have powerful effects on running time by themselves, the overall needs are derived from combinations of these effects. The highly complex problems now being studied in energy programs require computational models with higher dimensionality, with higher resolution, and with more physics.

HIGH-ENERGY AND NUCLEAR PHYSICS

The requirement for computers capable of meeting the data-reduction needs of a high-energy physics laboratory has been so great in the past that all other computing requirements could be met without significantly affecting the large central facility. However, in the decade of the 1980s, several new computational needs have appeared that require the unique capabilities of supercomputers and clearly require capabilities presently associated with Class VII systems.

The theoretical high-energy physics community represents an important class of users with very large computational needs. This is primarily the result of the rapid rise of computational quantum field theory, particularly in numerical studies of lattice-gauge theory. To put this development in perspective, it should be noted that computer simulation is a generic numerical tool for studying the behavior of particles and fields, and its importance does not rest on any particular fashion or on the currency of any particular theoretical idea. The ability to carry out such calculations is primarily a result of the rapid increase in available computer power, and, as such, it represents a permanent change in the way theoretical physics is done. The needs have fallen into two distinct categories. The first category includes the more traditional forms of such theoretical computations as numerical integration, solution of integral or differential equations, and calculations of Feynman diagrams. The second category is the large-scale numerical simulation of quantum field theory on a lattice. These calculations are highly central-processing-unit (CPU) intensive. The lattice-gauge-theory algorithms are relatively simple, repetitive, and easily vectorizable. Thus they are well suited to a variety of parallel and pipelined architectures provided that a large, faster-accessed memory is available. Even low-statistics calculations on modest-sized lattices require the equivalent of tens of Cray hours.

Two newly emerging needs for computer power beyond the scope of Class VI systems are from the accelerator- and experiment-design communities. An example of an accelerator-design requirement is for the turn-by-turn simulation of potential designs for the new superconducting super collider accelerator currently in conceptual design. The integrated time needs here are CPU times measured in Cray–1 equivalent years.

An example of the experiment-design-related requirement is the full simulation of Monte Carlo events in a colliding beam detector system. The number of simulated events run should, ideally, be substantially greater than the number of real physics events to be analyzed. Furthermore, since experimental results may change the way that a detector is tuned, it may be necessary to make the simulations concurrently with the taking of data, that is, when the data-reduction computers are most fully loaded.

Experimental high-energy physics data reduction, which has heretofore used standard general-purpose computers, also needs a new generation of computers. The generation of detectors now just coming into use necessarily gather data at very high rates in order to extract the physics of interest from the enormously large accompanying backgrounds. The volume of data collected from these new experiments is several orders of magnitude larger than in experiments performed in 1980. The computational problems are enormous and new classes of supercomputers along with the special-purpose processors appear to be the only practical way in which to satisfy these unfulfilled computational needs.

A computing center that would serve the interest of high-energy theorists should be endowed with one of the most powerful mainframes available, both in computational speed and in memory size, such as the Office of Energy Research Class VII system proposed for fiscal year 1988.

BASIC ENERGY SCIENCE

The development and proliferation of investigations of diverse material systems and phenomena via computer simulation and modeling is a rich field of scientific endeavour anchored in the physical sciences (with cross-fertilization links to advances in applied mathematics and computer science), made possible singularly by the advent of high-powered computers. Computer simulations provide information about phenomena and processes in material systems with refined microscopic spatial and temporal resolution and enable investigations of the dynamical evolution of complex systems under extreme conditions where data from experiments or other methods of investigation are not attainable. In addition such studies provide benchmarks for critical testing and refinement of theoretical concepts and aid in the interpretation of experimental observations.

Current simulation methods involve the generation and analysis of phase-space trajectories of an interacting many-particle system either by the direct numerical integration of the equations of motion—molecular dynamics (MD) and reaction-trajectory method—or via the sampling of phase-space configurations—Monte Carlo. In either case the many-body nature of the systems under study and the statistical modes of analyses dictate the necessity for extended computer time and storage capabilities.

The wide range of materials systems investigated by computer simulations include: the equilibrium and nonequilibrium structure and dynamics of materials at different states of aggregation (solids and liquids) and the kinetics and dynamics of phase transformations; properties of metastable systems (supercooled liquids, quenched liquids, gasses); homo- and multicomponent materials; ordered versus disordered (amorphous) solids; surfaces; interfaces and interphase interfaces, for example, solid-solid (superlattices and coherent structures), solid-liquid (epitaxial crystal growth and homogeneous nucleation), solid-gas (molecular beam epitaxy, heterogeneous catalysis).

Simulation studies on these systems allow investigation of structural and dynamical characteristics, kinetics and dynamics of phase-transformations, transport and nonlinear phenomena (heat, matter, electrical), diffusion processes, and reaction dynamics. Furthermore, modifications of the intrinsic properties of condensed matter systems and phenomena (such as fracture, solid transformations, and plastic flow), due to external fields (like mechanical stress and heat gradient) can be investigated. In addition to an improved understanding of existing material systems, simulation studies could serve as the impetus for exploration of methods of preparation and growth of novel materials.

Underlying simulation studies of extended condensed matter systems is the notion that the properties of the "calculational sample" on which the simulation is carried out, extended via the commonly used periodic boundary conditions, are a faithful representation of the nature of the macroscopic system. Among the factors that dictate the size of the calculational sample are the ranges of interparticle interaction potentials and fluctuation wavelengths. Thus, for example, the MD simulation of the structural and dynamical properties of a solid simple metal (e.g., A1) requires a system containing 2,000 particles; the simulation of binary liquid metals and supercooled liquids require an even larger sample due to concentration fluctuations. Simulations of stressed crystals, fracture and plastic flow, shock-wave propagation, the dynamics of melting, and hydrodynamical phenomena would require systems where the number of particles would be 5,000–10,000. In the presence of long-range and realistic multibody forces the computing time grows as a (low) power of the number of particles. Such requirements necessitate memory capacity beyond Cray–1 capability and large increases in computational speed.

A critical input in materials simulations is the interparticle interaction potential. A faithful simulation requires the calculation of such potentials via pseudo-potential methods that, for metallic systems, depend upon the thermodynamic state variables (density, temperature, pressure). Simulations of nonequilibrium phenomena (such as solidification and quenching) in which the state variables themselves evolve in time require a self-consistent adjustment of the interaction potentials along with the dynamical evolution of the system.

The coupled complexities of size and interaction potential calculations make such simulations prohibitive on the Cray–1. Furthermore, the magnitude of such simulations dictate substantial time requirements, for example, 80 minutes of Cray–1 time allow the generation of 5,000 integration time steps for a system containing 1,500 particles interacting via simple truncated Lennard-Jones potentials, with a fully optimized code. Note that this is the least demanding model from a computational point of view. A typical study of the solidification of such a system requires 50,000 integration time steps. It should be emphasized that the above considerations are dictated by the nature of the physical systems and phenomena and cannot be compromised by approximate treatments that will prejudice and distort the simulation results. Thus progress in this field can be made with substantial access to the Class VII computing facilities proposed for fiscal year 1988.

NOTE

1. "The Role of Supercomputers in Energy Research Programs," U.S. Department of Energy, DOE/ER–0218, February, 1985.

10

SUPERCOMPUTERS AT NASA

RANDOLPH GRAVES

This chapter reviews selected programs of NASA that relate to supercomputers and discusses the growing network resources being developed by the agency.

NASA is principally an agency of supercomputers. However, the agency does have an interest in advanced architectures and their impact on the types of problems NASA needs to solve. Accordingly, the agency has a small in-house and sponsored-research program associated with supercomputing. We have associated with NASA two research institutes that are administered by the University Space Research Association (USRA). Through its 54–university consortium USRA has access to a pool of expertise that can be called upon to assist in supercomputer R&D.

NASA R&D EFFORTS

USRA administers the Institute for Computer Applications and Science and Engineering (ICASE), which was founded in the early 1970s and is co-located at NASA's Langley Research Center. The principal focus of this institute is in applied mathematics and algorithm development for vector and parallel processing machines. ICASE has a minor focus in science and has assisted the Langley Research Center in a number of hardware and software research efforts including the development of the multiprocessor Finite Element Machine.

The second of our research institutions associated with supercomputing is the Research Institute for Advanced Computer Science (RIACS), which is co-located at our NASA Ames facility at Moffett Field. The principal focus of RIACS is computer science and its principal research area is in algorithm development and advanced computer architectures and systems software for concurrent processing systems and their application to aeronautics and space activities.

The second area of interest at RIACS is in expert systems, the use of artificial intelligence insofar as it affects supercomputer usage. NASA is looking at expert systems as an expert or intelligent interface between the user and the supercomputer. While this project is currently a small effort, it is one we expect to grow steadily.

In addition to its in-house and research-institute activities, NASA also sponsors research and development efforts through grants and contracts. The agency supports architectural feasibility studies, and occasionally funds the development of supercomputers for specific applications, such as the parallel processor computer located at the Goddard Spaceflight Center, which is used for processing satellite images.

NASA INVOLVEMENT WITH SUPERCOMPUTERS

NASA's principal involvement with supercomputers is a program begun in the mid 1970s to establish a major national supercomputer center for aeronautical R&D. This program is termed the "numerical aerodynamic simulation" (NAS) program, and it is aimed at producing a national facility that will be available to the entire computational fluid dynamics community for pathfinding research in aerodynamics and aerothermodynamics.

In September, 1985, NASA took delivery of a Cray–2 computer system, which at that time was the largest supercomputer in existence, with 256 million words of memory and four processors. The Cray–2 is supported by a processing subsystem consisting of two Amdahl 5840 computers that serve as the "front end" support for the NAS processing system.

A key element for remote access is in the communication system. NASA has recently signed a major contract with the Boeing Computer Services to institute an agencywide communications system that is principally satellite based, but also includes land lines. The communications network will be initially a 1.44 megabit per second link that will interconnect three of the NASA research centers and later include more NASA centers, as well as off-site contractors and selected universities. NASA is currently negotiating with Boeing Computer Services to upgrade the data communications capability to six megabits so that the network can handle interactive computing from remote stations. The NASA centers to which satellite links are connected will serve node points for land lines that will go out from the NASA centers to contractors and universities performing computational research and applications under grant or contract.

The NASA program projects a major need in the area of mass storage. Initially there will be a relatively large disk farm associated with the facility. NASA is currently supporting research with which other federal agencies are assisting in the development of an erasable optical disk, a unit that would have a trillion bits of storage. The unit would have a transfer capacity of over a gigabit in each direction (in and out) simultaneously. For on-line supercomputer applications the optical disk driver is being referred to as an optical buffer. It would replace between 200 and 300 normal magnetic disk drives. NASA expects a prototype capability to be available sometime in late 1987 with commercial units available at a later date.

In early 1986 the NAS Projects Office issued a request-for-proposal for a high-speed processor to be brought into the NAS system in early 1987 after completion of the new NAS building. Once the second high-speed processor is installed, completion of the NASA facility development is anticipated in 1989 and should be fully operational in all aspects.

Table 10.1
Computational Aerodynamics

DEFINITION:

- Simulation of aerodynamic flows through the numerical solution of approximating sets of fluid dynamic equations using high-speed computers

GOAL:

- Obtain by computation aerodynamic information normally measured in wind tunnels, flight tests or other ground-based facilities

MAJOR BENEFITS:

- Significantly improved designs resulting from (1) use of numerical optimization procedures and (2) simulation of full-scale free-flight conditions

- Increased efficiency of wind-tunnel testing

- Reduced design time, cost and risk for aerospace vehicle development

---

Source: All tables and figures in this chapter are from the National Aeronautics and Space Administration.

Table 10.2
Examples of Important Viscous-Dominated Problems

- Prediction of Aerodynamic forces

- Inlet flows

- Compressor stall

- Airframe/Propulsion system integration

- Strake design

- Stall/Buffet

- Performance Near Performance Boundaries

Figure 10.1
Revolutionary Advances in Inviscid-Flow Technology During the 1970s

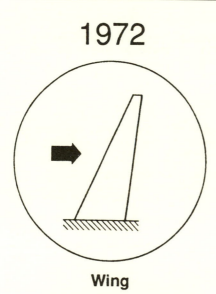

1972

Wing
18 Hours of Computation

1980

Complete Aircraft
15 Minutes of Computation

Figure 10.2
Governing Equations and Computer Requirements for Computational Aerodynamics

APPROXIMATION	CAPABILITY	GRID POINTS REQUIRED	COMPUTER REQUIREMENTS
Linearized Inviscid	Subsonic/Supersonic Pressure Loads Vortex Drag	3×10^3 Panels	1/10 Class VI
Nonlinear Inviscid	Above Plus: Transonic Pressure Loads Wave Drag	10^5	Class VI
Reynolds Averaged Navier Stokes	Above Plus: Separation/Reattachment Stall/Buffet/Flutter Total Drag	10^7	30 X Class VI
Large Eddy Simulation	Above Plus: Turbulence Structure Aerodynamic Noise	10^9	3000 X Class VI
Full Navier Stokes	Above Plus: Laminar/Turbulent Transition Turbulence Dissipation	10^{12} to 10^{15}	3 Million to 3 Billion Class VI

Figure 10.3
Computer Speed and Memory Requirements Compared with Computer Capabilities

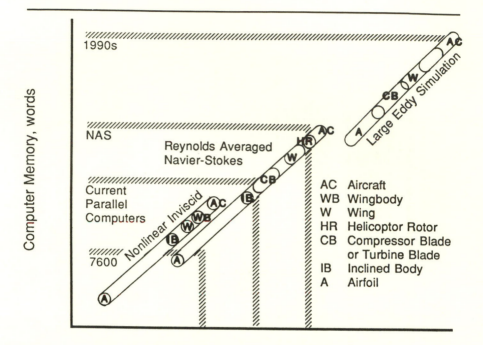

Computer Speed, M Flops

Table 10.3
Key Technical Goals

- INITIAL OPERATING CONFIGURATION (IOC) IN 1986

  - High-speed processor-1 (HSP-1)
    + 4 to 6 times current supercomputer performance
  - Integrated and expandable configuration
  - Uniform user interface to all subsystems (UNIX environment)
  - Wideband communications to other NASA centers

- EXTENDED OPERATING CONFIGURATION (EOC) IN 1988

  - Additional high-speed processor (HSP-2)
    + 16 to 24 times current supercomputer performance
  - Upgraded subsystems and graphics capabilities
  - Wideband communications to national user community
  - DOD classified/secure operations (not NASA funded)

- FURTHER EXTENSIONS IN EARLY 1990s

  - Replace high-speed processor-1 wit advanced supercomputer
    + 100 times current supercomputer performance

Figure 10.4
NASA Operations Schedule

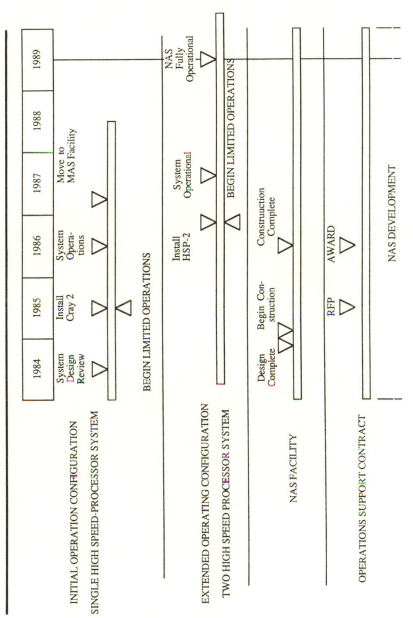

135

Table 10.4
Summary

The Goal of the NAS program is to optimize the computational
process from problem formulation to publication of results

The NAS will be developed in an on-going evolutionary fashion
using commercially available hardware and software supplied
by multiple vendors

The NAS will be a NASA-managed jointly used national resource

Computational aerodynamics is becoming a vital tool for
aircraft design and the NAS is needed to assure a continued
leadership position in its development

Where do we go from here? 1989 will not be the end of the road. For
example, around 1976 NASA indicated its need for a machine that had
240 million words of memory and a gigaflop performance in order to cal-
culate the fluid flow about a complete aircraft. At that time, the computer
manufacturers said such a machine was not possible. NASA kept pushing, through programs like NAS, and now we are approximately there.
The next step, of course, is in the 1990s. NASA's aeronautics and space
R & D programs are looking for a machine with at least a trillion floating-
point operations per second and several hundred billion words of memory. Through scientific exchange and linkage between government, universities, industrial users, and computer manufacturers it is anticipated
that this objective can be met before the mid-1990s.

SUMMARY

Tables 10.1–10.4 and Figures 10.1–10.4 summarize advances in NA-
SA's Numerical Aerodynamic Simulation Program. Table 10.1 indicates
definition, goal, and benefit of supercomputer applications to computational aerodynamics. Table 10.2 provides examples of viscous-dominated problems related to aerodynamic design. Figure 10.1 reflects the
time-cost saving benefits of advances in inviscid-flow technology computation. Figure 10.2 lists equations and computer requirements for computational aerodynamics. Figure 10.3 illustrates relationships between
computer memory and computer speed when applied to aerodynamic
design problems. Table 10.3 summarizes key technical goals in configuration requirements for supercomputers. Figure 10.4 outlines the operational schedule for the NASA program. Table 10.4 summarizes key
objectives and functions of the NASA program.

11

NATIONAL BUREAU OF STANDARDS: OUTREACH TO THE SUPERCOMPUTER COMMUNITY

GLENN INGRAM

OVERVIEW OF THE NATIONAL BUREAU OF STANDARDS

The National Bureau of Standards was created by Congress in 1901 to meet the needs of the United States for a unified system of physical, chemical, and engineering measurements. Its basic mission is to provide for the nation's measurements and standards needs, and additional legislative directives through the years have broadened the scope of NBS programs.

NBS pursues its mission through service as a central reference laboratory and lead agency for measurements and standards, and through research and development activities in various areas of science and engineering. This has led to extensive interactions with both industry and universities: There are about 150 industrial research associates, and about 700 visiting faculty and students at NBS.

Some specialized facilities are a focus for collaborations, and NBS provides various services in calibrations and standard reference material. Also, there are some 500 contracts with other agencies.

SUPERCOMPUTING AT NBS

A system consisting of a CDC CYBER 855 and 205 supercomputer was accepted by NBS in early May 1983. This new facility is the outcome of a competitive procurement, and results from actions by the Department of Commerce to combine the computing requirements of NBS and the Environmental Research Laboratories (ERL) of the National Oceanic and Atmospheric Administration. This system will serve both NBS sites (Gaithersburg, MD, and Boulder, CO) five ERL sites (Boulder, CO; Seattle, WA; Ann Arbor, MI; Norman, OK; and Miami, FL), and the National Telecommunications and Information Agency's Institute for Telecommunications Sciences (ITS) in Boulder.

At this time, interim communications networks link these sites. The long-term network will be operational in the fall of 1983, with communications between VAX II/780s at each ERL site (plus a medium-scale computer in Boulder) and the supercomputer system. Remote and local users have interactive access to the 855, while the 205 is used in batch mode.

No classified work will be done in this facility, and it is planned to provide service to a variety of scientific and engineering areas, including an emphasis on atmospheric and oceanographic modeling at the ERL sites.

POTENTIAL FOR THE NBS CONTRIBUTIONS IN SUPERCOMPUTING

A unique feature of the organization of computing at NBS is the location of central computing functions in the Center for Applied Mathematics. This has fostered a very close and effective interaction between mathematical disciplines and applied computer science.

With the creation of the Scientific Computing Division seven years ago, there was an emphasis on recruiting staff in computational mathematics—for research, consulting, and software. Significant strength has developed in this area; in addition to research and consulting, this staff has assembled the Core Mathematical Library—a collection of the best nonproprietary mathematical routines from a variety of sources. It also developed a Guide to Available Mathematical Software, a taxonomy of software that is dynamic and expandable. It includes proprietary packages and three levels of documentation. There is an on-line version that will continue to evolve in providing aid to users in the choice of routines.

The long-standing pattern of interactions with both universities and industry provides the basis for a special role for NBS in supercomputing. This history is ideally suited to serving as a catalyst for collaboration among various organizations. At present, there are cooperative agreements with the supercomputer facilities at Colorado State University and the University of Minnesota. We are interested in expanding such contracts.

Building on this background, NBS offered a workshop, cosponsored by SIAM, on scientific software for supercomputing in October 1986. It included such topics as available scientific software, algorithm design, languages, vector and parallel architecture, and case studies, and provided for hands-on experience.

The supercomputer era is still quite new at NBS; staff members have used such facilities at other sites for a number of years, but now we are coping with the problems of staffing, managing, and operating this system and the network that links it to remote sites. After one month, use patterns are not clear and conversions from other computers are in process. Use of the CYBER 855/205 system by anyone outside the three basic partners (NBS, ERL, and ITS) is limited to those who are involved in active collaboration with staff at one of those three agencies.

RESEARCH AND DEVELOPMENT NEEDS

The tremendous advances in supercomputer hardware serve to highlight other areas in which major improvements are needed. Any attempt to categorize these needs encounters inevitable overlap, but there are technological, software, and mathematical aspects.

One clear requirement is for effective, reliable, and very-large-capacity mass storage systems. The power of the supercomputer and dispersed user communities significantly increase this requirement. Devices that have been and are being used have played an important role in several computing centers, but an order-of-magnitude increase in capacity is needed now.

Computers that span the range from micro to super are available. Using this range of capabilities in a complementary, cost-effective way involves very thoughtful planning and continuing research and development in telecommunications, software, graphics, and standardization.

In scientific software, there is a need for continuing developmental efforts on mathematical, statistical, and graphic programs, with special attention to compatibility and portability. Closely associated with this is the continued development of help and documentation systems to guide users in the choice of mathematical routines. Development of preprocessors for supercomputer compilers for specific classes of applications can have a major effect on performance.

Finally, more work should be supported in algorithm design for vector and parallel processors.

OTHER ACTIONS

Supercomputers have created additional needs for training, which can be described in two categories.

There is a severe shortage of people with experience in supercomputer centers and real competition for them. Hence, every new supercomputer facility needs training for its existing staff; some of this can be provided by vendors, but it serves simply as an introduction. Increased interactions among computing centers that focus on training for staff could be very valuable to new facilities.

For example, if an institution planning to acquire a supercomputer could send some staff members to an experienced site, this staff could acquire training by immersion. That is, they would be embedded in the facility with more experienced staff and learn on the job. To be most effective, the existing supercomputer site should have approximately the same equipment and the same operating system that will be acquired by the new center. Another possibility is to have a mature supercomputer center loan two or three of its staff to a new installation during its start-up phase. Of course, both of these approaches could be used, and certainly this would do a lot to improve the start of a new facility.

At another level, there is a need to provide training to users—typically scientists and engineers. This, too, calls for collaboration between established and beginning sites. One obvious action is to promote visits to institutions with young supercomputers by software and algorithm experts. Such visits could include seminars and demonstrations. A complementary activity would be to promote longer visits to institutions with mature facilities by faculty, research staff, and graduate students.

Another significant effort is illustrated by the workshop to be sponsored by NBS and SIAM. Such workshops should be held where there is access to a supercomputer and where specific high-quality scientific software is available. Presentations of case studies that are not research reports on the results of computation but, rather, careful descriptions of model development, could be very helpful.

PART IV
SUPERCOMPUTERS AND THE UNIVERSITY

INTRODUCTION

Universities have emerged as a key element in the evolving federal policies that impact supercomputers. Responding to recommendations by the Lax Committee and the House Science and Technology Committee, several universities bought advanced machines and established supercomputer centers early in the decade. These universities, including Minnesota, Colorado State, and Purdue, led the way in providing supercomputer access for academic scientists and their students and to industry as supercomputer technology began to find applications in the aerospace, energy, and automotive industries. Spurred by the need for more scientists with experience in supercomputer technology and the need for training and testing sites for manufacturers of supercomputers, other universities developed proposals, generated funding for centers, and established training and research institutes.

The chapters in this section describe the various missions, structures, and activities of a cross section of university-based supercomputer centers. Each has a unique approach and offers a different structure for both research and training.

The three National Science Foundation centers are linked together by the common thread of NSF funding and the developing communications network between the centers. Nevertheless, each is organized differently and has carved out its own mission within the parameters of the NSF grant.

Florida State University, which hosted the June 1985 conference on which this volume is based, is a center funded by a coalition of federal, state, business, and university resources. Forbidden by state law to undertake classified research, this center concentrates on training, basic research in energy-related areas, and theory.

The University of Georgia is unique because the center in Athens is funded solely by the State of Georgia. This center serves a number of state needs, provides supercomputer facilities for its own faculty and for that of The Georgia Institute of Technology, and seeks contract research from both government and industry.

Collectively, these university-based supercomputers and the centers they support should develop into a major national resource to provide basic research, trained scientists, and technological resources for industry as well as serve as the linchpin of U.S. comprehensive national security.

12

NATIONAL SCIENCE FOUNDATION CENTERS
THE CORNELL APPROACH TO TRAINING AND RESEARCH

KEN WILSON

Cornell University has established a national supercomputer research center through a grant provided by the Advanced Scientific Computing Initiative of the National Science Foundation. It is too early to see major results from the initiative. Cornell has been operating an interim supercomputing system since May 1, 1985, but facilities are still primitive and error-prone; the full program at Cornell and at the other centers will not be under way until the fall or winter. However, there are a number of general comments and concerns that are of importance.

HIGH-SPEED NETWORKING CRITICAL

High-speed networking on a national scale remains critical to the success of the initiative. The appointment of Dennis Jennings to head the networking effort at the NSF has laid the basis for progress in the networking area, but continued attention and funding will be needed for progress. High-quality graphics displays are becoming essential tools for researchers using supercomputers; as such graphics displays become available to supercomputer users around the country, the pressure for high-speed networking will grow very rapidly.

Both technically and politically, regional networks at 1 megabaud rates (T1 lines) are now becoming feasible, and the best route to networking appears to be regional networks established through local initiatives (like the Princeton consortium network, and the North Carolina and Michigan State networks). These regional networks can then be linked by the NSF networking program.

Supercomputers are unique instruments of science in their breadth. Every scientific discipline, every engineering discipline, and many areas of medicine, agriculture, social sciences, and architecture require supercomputer access now or in the future. Funding for the centers is meager when this breadth is considered. For example, Cornell University will receive roughly $7 million per year from the National Science Foundation to operate one of the four national centers. The total sponsored research budget for Cornell's graduate programs is over $170 million per year; $7 million more could easily be justified in order to meet Cornell's supercomputing needs alone. By comparison, a major telescope, an accelerator, or a magnet facility serves only one discipline or, at most, a very small fraction of the disciplines a supercomputer can serve.

BASIC RESEARCH LEADS TO SPIN-OFFS

Basic research is characterized by the search for far-reaching breakthroughs in understanding nature, which have many unexpected practical spin-offs and which also lay the basis for further research advances. This is especially true of supercomputing. Much of the research using supercomputers will be concerned with applying the basic laws of nature to real-life problems. There are only a handful of these laws, such as Newton's laws of mechanical motion, Maxwell's laws for electromagnetic phenomena, and Schrodinger's equation for automatic and molecular phenomena. These precise laws are reliable as the rising and setting of the sun. They are applicable to an extraordinary variety of circumstances. For example, every metal, every biological molecule up to and including DNA, and every exotic material (such as carbon fibers) has properties completely determined by the Schrodinger equation. The problem facing researchers is how to unlock the secrets of these equations in countless practical situations. Supercomputers are a critically needed tool for this research but breakthroughs in understanding how to use supercomputers will also be required. The most-rewarding research breakthroughs will be those that lead to progress in many different application areas at once—from weather forecasting to drug design. Such breakthroughs will result from increased understanding of the laws of nature and how to work with them effectively.

Researchers often pose sample problems to be solved, by themselves or by their graduate students, which illustrate basic features of the laws of nature but do not have much immediate practical value. The understanding that results from solving these problems can have practical benefits. For example, when I was a graduate student I studied problems at the far-out fringes of elementary particle physics; even my closest colleagues found it difficult to justify what I was doing. During this time, I applied for a Bell Laboratories Fellowship and stated on the application that my research plans bore no relation whatsoever to the needs of the telephone company. I was mistaken. Although I had no way of knowing at the time, ten years later my research had progressed in unexpected directions involving properties of matter that had become of very great interest to physicists at Bell Laboratories. Even today, almost 30 years and one Nobel Prize later, it is not possible to predict where the most important spin-offs of my research will occur: in oil exploration? the design of new materials? social or economic studies?

I wish to emphasize the importance of not targeting basic research to specific applications. We need basic research in all of its generality leading to spin-offs throughout the civilian and military sector—to improve our economic competitiveness, to improve our health, and to advance our defenses. More attention needs to be given to the spin-off process, but targeting basic research is not the answer. For example, the deputy director of the Cornell Center, Ravi Sudan, is also director of Cornell's Plasma Physics Center and he is now struggling with cutbacks in the plasma fusion program; his research is basic and should be considered part of the U.S. basic research effort rather than as specifically tied to the fortunes of the thermonuclear program. Professor Sudan's background in plasma physics will be very helpful in dealing with many different kinds of fluid and gas flow problems that will be studied on Cornell's supercomputer. It is especially important that basic research not be tied politically to the success of specific development projects like the Strategic Defense Initiative. Instead, the United States should encourage basic research to pursue its traditional search for generality, and simultaneously improve the spin-off process so that important research breakthroughs are milked for all kinds of spin-offs rather than being locked into one development project.

One important trend that can improve the research spin-off process is the growth of the graduate engineering profession. Doctorate-level engineers are better prepared to help move university research results into practice than engineers with just an undergraduate degree. At Cornell, the engineering college has changed over the last 20 years from a predominantly undergraduate program to a balanced mix of undergraduate and graduate training.

It is important that U.S. civilian manufacturing industries be provided their fair share of access to Ph.D. talent. Success in international competition with its resulting effects on both the national debt and the trade deficit depends increasingly on using high technology, including supercomputers and university research results, to design more competitive products for the world market. Doctorate-level scientists and engineers are needed to help civilian industries take advantage of high technology opportunities to improve design and manufacturing.

PARALLELISM KEY TO ADVANCES

The 1990s could bring vastly more powerful supercomputers than any that are available or are even currently talked about. The key to revolutionary advances is through the use of parallelism: computers that can execute thousands or even millions of operations simultaneously. Parallelism on the scale of thousands of operations has been amply demonstrated in an academic setting, notably at the University of Edinburgh in Scotland, which has two ICL distributed array processors, each capable of executing 4,000 operations simultaneously. The main need currently is to make available and win commercial success for highly parallel computing systems with computing power vastly beyond today's Cray, CYBER, and Japanese supercomputers. Cornell is pressing the U.S. computer industry to provide such a system to Cornell within the next year or two.

Each of the university-based supercomputer centers is likely to develop its own areas of special expertise. Cornell is largely focused on the future of supercomputing. It already has projects under way to address bottlenecks that could prevent parallel machines from being both academically and commercially successful. Cornell has projects in software productivity, very advanced graphics, parallel processing systems, and key application areas for parallel processing.

The most important area of all for the future of very powerful systems is that of atomic and molecular physics with applications to properties and design of materials. All present supercomputers are hopelessly inadequate for solving such problems; if one learns how to solve these problems on future machines, the payoff could be spectacular. Experimental physics, chemistry, and biology have not even scratched the surface of the totality of chemical and material substances that could be industrially important; breakthroughs in the understanding and design of materials could lead to whole new industrial groupings alongside the aluminum, silicon, glass, and oil industries, each of which has a single material as its base. Cornell is the principal center with management trained to conduct and assess research in this area and, in fact, I have a joint research effort with John Wilkins of Cornell trying to analyze and improve on some promising new methods for solving the Schrodinger equation for atoms and molecules.

At present the supercomputer market is small; supercomputers are built largely by hand at enormous cost. Meanwhile, the need for supercomputing in private industry and universities is growing explosively. Unfortunately, the $5 million entry-level price of today's supercomputers puts them out of reach of all but the largest corporations. Even in Fortune 500 companies, the process of deciding to buy a multi-million-dollar supercomputer requires three years of intense battle between groups of scientists and engineers who want it versus company accountants who demand, but do not get, a yearly payback schedule for the investment. Most industrial scientists and engineers prefer to avoid this battle and opt instead to buy much less powerful, cheaper superminicomputers. The result of this has been twofold. First, Digital Equipment Corporation has become the second largest computer company in the United States with several billion dollars a year in superminicomputer sales, while supercomputer sales are far lower—only a couple hundred million dollars per year. The second consequence is that many U.S. industrial scientists and engineers are very poorly equipped to deal with today's intense international competition. Their computing jobs may take days or weeks to run on their superminicomputers and in this case they desperately need the faster turnaround that supercomputers could provide.

Through parallel computer design it is now feasible to produce new supercomputer product lines that would have an entry-level price below $100,000 and then would be upgradeable in $100,000 increments all the way to $100 million dollar behemoths vastly surpassing today's supercomputers. Such supercomputers could be mass-produced at low cost using today's advanced but mass-produced VLSI silicon chips. Given proper support, they would enjoy a huge market from U.S. private industry, which needs to migrate en masse from their present inadequate systems.

SUPPORT NEEDED TO OVERCOME LAG

Unfortunately, the proviso "with proper support" is a key sticking point. Private industry has accumulated a massive amount of software targeted to its existing computers, none of which can easily be moved to the revolutionary new parallel systems. Such systems have to accumulate both systems support and new applications starting from scratch, with almost a 20–year lag behind today's already mature computers. A massive effort will be required to overcome this 20–year lag. Cornell is trying to build a coalition of adequate size to deal with this problem, including many computing manufacturers, national laboratories, and about 30 leading corporate users of scientific computing. The key to our program is the coalition of industrial users who define the ultimate market for new systems. It is taking time to build an Industrial Associates Program at Cornell, but we continue to work hard. Many universities have computer-science projects related to parallel processing that will help our efforts; as far as we know, Cornell is the only university focusing on helping the commercialization of parallel processing so that its benefits will become very widely available to both universities and industry.

THE ILLINOIS SYSTEM

LARRY SMARR

A bold initiative taken in 1984 by the Committee on Science and Technology of the U.S. House of Representatives and the National Science Foundation led to the establishment of the National Center for Supercomputing Applications (NCSA) at the University of Illinois at Urbana-Champaign. This new center, one of the largest projects in the history of the University of Illinois, is jointly funded by the Office of Advanced Scientific Computing at the National Science Foundation, the University of Illinois, the State of Illinois, and Cray Research, Inc.

THE ILLINOIS CENTER

The Illinois Center was designed in a manner at once both very conservative and quite radical. This is because the center is intended to deliver reliable service in the shortest possible time, while at the same time developing new ways to make supercomputers much more accessible to the scientific community.

The major components of the facility are based on the existing strength of the hardware and software that have been developed with thousands of man-years of effort at the Department of Energy national laboratories. The center will use the Cray Research, Inc. line of supercomputers. This arrangement will allow the center to run the wide universe of software for the Cray that has been developed over the last ten years. This is a very conservative approach, because without a widely available variety of debugged and reliable software, a fast machine is really of very little use to most audiences.

The operating system will be the Cray Time-Sharing System, which was developed at the DOE national labs, and which allows a number of users to work on the Cray supercomputer simultaneously. There are a wide variety of software productivity tools available with this operating system. The center will use the Common File System, developed by Los Alamos National Laboratory, as the software that keeps track of user files. Using CFS, scientists can within seconds select a file from up to one terabit (one thousand billion bits) of information available on the mass storage unit.

RAPID SUPERCOMPUTER UPGRADING

Although the choice of supercomputer is conservative, the University of Illinois has pioneered the radical concept of rapid upgrades in the power of that supercomputer. This is possible because we have entered the age of the "multiprocessor" supercomputer. This is a supercomputer, composed of several identical computers, that "gang tackles" a problem. This concept evolved from the innovative ILLIAC IV multiprocessor supercomputer designed at the University of Illinois in the 1960s. While all processors of the ILLIAC IV had to work in unison, today's multiprocessors have each processor independently work on a separate part of the problem. Illinois will start with the Cray XMP 124 multiprocessor, which contains two supercomputers, each more powerful than the previous Cray–1 single processor supercomputer.

The center will upgrade its supercomputer every one to two years. Thus, the machines will increase from 4 to 8, and finally to 16 processors at the end of the five-year NSF agreement. Because each individual processor also gets faster in each new generation, in five years the Illinois center will have moved to a machine that is 50 to 100 times faster than the current Cray–1 supercomputer.

All future, fast machines must have multiprocessor architecture if they are to be competitive. Unfortunately, almost all computer programs are written for single-processor computers. Therefore it is crucial that the basic research community learn immediately how to use multiprocessors.

The University of Illinois has established a second supercomputer center, the Center for Supercomputing Research and Development (CSRD), whose mission is to explore the computer science of these new problems. That center will be directed by Professor David Kuck, a pioneer of supercomputer architecture and software. The center will build an experimental multiprocessor, develop new software and operating systems for multiprocessors, and provide tools to restructure old FORTRAN computer programs to run on these new machines. The NCSA will work closely with the CSRD at Illinois to help transfer these new ideas to the national basic research community.

The University of Illinois is unique in having two supercomputer centers. NCSA will take the best hardware and software from the marketplace and the government labs and make it available to the national basic research community. CSRD will experimentally build prototypes of the supercomputers and their software, which will become commercially available in the 1990s and beyond.

IMPROVING ACCESS TO SUPERCOMPUTERS

While enormous computational power is becoming available to researchers, supercomputers are unbalanced with respect to delivering information to a human being. A good analogy is that of a large water reservoir near a large city. When the spillway is opened, an enormous volume of water is released. This must be channeled through a series of large distribution pipes to various parts of the city, then smaller pipes deliver it to neighborhoods. Within the neighborhoods smaller pipes deliver it to houses, and finally in a house an individual can turn on a faucet and get one or two drops of water at a time. The flow of information from a supercomputer is similar. Unfortunately, the United States has very little other than the large spillway available. The University of Illinois is trying to help develop a distribution system for this information so that a single user can get a "few drops" of information at a time.

The key to success in this area has two parts. First is the personal computer (PC) revolution. This rather unexpected development of technology has perfected the "faucet" of our delivery system. The PC is the device that all of us, whether we are at home or at work or in the laboratory, have had to learn to use to retrieve information. As such, researchers have become familiar with the "user-friendly" interface on their PCs. At Illinois, scientists are undertaking research to develop software to run the Cray from personal computers in such a way that a user may think the Cray is "inside the PC." The ideal is that within a few years a researcher will be able to plug his or her lap-top PC into a telephone anywhere in the country and run the Cray at the University of Illinois. This will greatly enhance scientists' access to supercomputers.

The second major part of the distribution system that must be developed is that of the network. The Illinois center is striving to develop a large range of data transmission speeds, in analogy to the large range of diameters of pipe for the water distribution system, so that the appropriate technology can be used at the appropriate place in the distribution system. For each of these transmission speeds there is an appropriate personal computer or scientific work station that can handle the information flow and perform tasks from simple text processing at the low end to elaborate three-dimensional color-simulation movies at the high end. This coupling of the personal computer revolution to the supercomputer revolution is of utmost importance if the scientific community will make the most productive use of these scarce resources.

Congress must understand the enormity of the task in creating a national network, which hooks all researchers together at high speed, including the Illinois center, other universities, national labs, and private companies. The only part of this network for which the Illinois center is funded, besides development and experiments on networking, is to set up a "front end" computer at Illinois to use existing telephone lines and to become part of the Department of Defense's ARPANET. Congress should appropriate sufficient funds to make the National Science Foundation's plans for a basic research national network a reality. However, the effort will be a major one lasting many years, just as have been the previous networking efforts, which joined this country together with roads, telephones, television, and electrical power lines.

National policy placed supercomputer centers in some of the great research universities of this country, so that we can begin research projects while the national network is developed. This staged process will allow for the training of thousands of new students on these new technologies. These students will go to other universities, to industry, and to government laboratories carrying skills and new approaches that will hasten the transition to supercomputing in their new jobs.

Merely delivering supercomputer cycles and training students is not an adequate response to the supercomputer challenge. Today's methods of using supercomputers in research are not nearly productive enough. All parts of the computer revolution must be linked together to increase our scientists' ability to realize the value of each supercomputer project.

A MULTIDISCIPLINARY INTELLECTUAL CENTER

Therefore, as part of Illinois's cost-sharing we have established an Interdisciplinary Research Center at the hub of NCSA. This center is located in the heart of the campus to foster research by scientists, engineers, social scientists, scholars, and computer professionals in a new type of scientific institute. This institute is multidisciplinary, with a mission to seek out common problems researchers experience in using common computational tools. These researchers will work in a modern comprehensive computational environment. Each desk will have a personal computer or a scientific work station networked to the supercomputer. Laser printers will produce typeset-quality output. Users will be able to view color movies of their simulations directly on their work station. Electronic mail will link all researchers in a common dialogue. The goal of the institute is to experimentally determine how much scientific productivity can be improved by removing the technological bottlenecks that have always hindered our rate of progress.

Researchers working side by side with nationally selected computer professionals in this state-of-the-art computational environment should be able to find innovative solutions to many of these problems. The institute will have a strong national visitors program to disseminate these new tools to the research community. The university-based supercomputer centers can have a major impact on the competitiveness of U.S. industry. Although most corporations know this, they are not in the business of performing the type of experimental, basic research that the centers can. Therefore, the centers may become the nucleus for a new type of partnership among government, universities, and industry to assure that these technologies can be developed and applied to American industries in time to allow them to remain competitive in the global marketplace.

ENHANCEMENT OF THE CENTER

The Illinois center is pursuing discussions with research corporations to see if a major enhancement of NCSA would be possible. These corporate partners will not replace the NSF dollars for university basic research. Rather, they can augment the existing program in ways that help transfer the new technologies directly to industry. The enhancements in scientific productivity developed in university centers should translate into engineering and manufacturing productivities in the marketplace. This transfer will not happen without a new framework for transferring technology. The university centers can serve as the neutral ground for this transfer by developing human partnerships between research scientists in academia and in industry to work together in the expanded centers.

The governor and General Assembly of Illinois and the University of Illinois have contributed to the building of this center some $21.5 million over the five years of this program. In addition, Cray Research is supplying some $8 million, and the National Science Foundation's share is $43.9 million, for a total of $73.4 million over five years. Thus, the State of Illinois and Cray Research are supporting 40 percent of the center's program, with NSF supporting the remaining 60 percent. This partnership among federal, state, and industrial sources together with the university is what is making the national centers possible.

However, to build and sustain a world-class center, the five-year commitment by NSF must be met and continued beyond the initial five years. These centers, if successful, should provide technological leadership well into the next century. It would be a tremendous waste of the funds already invested if the centers, once started, were starved for the necessary funds to carry out their missions.

It is important to realize that there is an enormous economy-of-scale to be had by augmenting the computational resources at the established centers rather than creating new centers. To add a second supercomputer is a fairly marginal cost to an operational center; funds must be found for the second machine and the incremental staff needed to run it, but these costs are a small fraction of the cost of establishing a new center from the ground up. Therefore it is probably more sensible to follow a two-track goal to increase national supercomputer capacity. The first track would place more of the latest supercomputers in the existing centers as saturation occurs. The second track would place the new "minisupercomputers," which cost under $1 million each, in remote university research groups that have demonstrated need for large amounts of production time.

SUMMARY

Congress and NSF have started a bold initiative that will have far-reaching consequences for the basic research effort of the United States. Furthermore, there is a promise for revitalizing U.S. industry if corporations join with the partnership already formed between the federal and state governments, universities, and the computer vendors. The key factor to be considered is that this is a long-term partnership that is going to require increasing investment for some years to come.

Assuming we can all keep the vision of coupling the computer and scientific revolutions, which brought us together, I am confident of undreamed-of results from this program.

THE SAN DIEGO SUPERCOMPUTER CENTER

SIDNEY KARIN

The San Diego Supercomputer Center (SDSC), located on the campus of the University of California, San Diego, is administered and operated by GA Technologies under contract to the National Science Foundation. We began providing user services early in 1986.

GA is a research company, privately owned, but operating in a mode similar to the national labs. We have been involved in fusion research for many years, and have one of the largest magnetic fusion experiments in the world. In the course of that research, we have been connected to the National Magnetic Fusion Energy Computer Center Network (MFE-NET) for several years. MFENET uses supercomputers located at the Lawrence Livermore National Laboratory, and links the research that we are doing in San Diego with other fusion researchers located across the nation—at Princeton University, MIT, and the University of Texas, for example. The daily use of the latest scientific computational resources to conduct actual research has built a base of practical experience at GA that enhances the physical operation of the SDSC, enabling us to assist users in taking full advantage of the powerful resources available to them.

The users of the San Diego Supercomputer Center are scientists associated with some of the most distinguished research institutions in the world, from Hawaii in the Pacific to the East Coast of the continental United States. Members of the consortium that provides policy guidance to the center include:

Agouron Institute

California Institute of Technology

National Optical Astronomy Observatories

Research Institute of Scripps Clinic

Salk Institute for Biological Studies

San Diego State University

Scripps Institution of Oceanography

Southwest Fisheries Center

Stanford University

University of Southern California

University of California at Berkeley

University of California at Davis

University of California at Irvine

University of California at Los Angeles

University of California at Riverside

University of California at San Diego

University of California at San Francisco

University of California at Santa Cruz

University of Hawaii

University of Maryland

University of Michigan

University of Utah

University of Washington

University of Wisconsin

Most of these institutions are already connected to SDSC via remote user access centers (see Figure 12.1) located on their campuses and via high-speed communications systems; all will eventually have such connections. Links to the other national centers and a higher-speed communications system are now being put in place.

REMOTE USER SERVICES

Users of our center have available—working from their own desk-top work stations—the most powerful computational resources available today through a network functioning identically to MFENET. The software used for this network allows as many as 200 users at a time to access the computer from their own work stations just as if the computer were located right next door. In addition to bringing the interactive operating system to a national community of scientists and engineers, SDSCNET also permits direct communications among geographically separated researchers, encouraging collaboration of scientists across the nation studying similar problems. SDSC is currently implementing support for

Figure 12.1
Typical Remote User Access Center

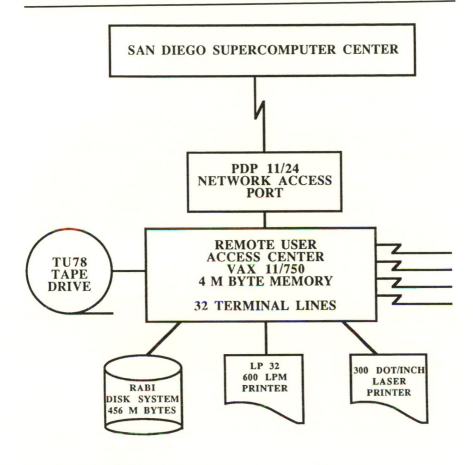

Source: Compiled by the author.

TCP/IP-based communications and is integrating the SDSCNET with the emerging NSFNET.

The network and high-speed communications systems put the center's complete resources at the disposal of researchers at our consortium institutions, making it possible to study far more complex problems than in the past. It also speeds up the whole experimentation process. Because it is interactive and because data can be transmitted very rapidly, the scientist gets immediate results, permitting, for example, numerous experimental modeling opportunities within the time period normally consumed waiting for one experimental result using batch systems. This feature is key to obtaining the maximum scientific research benefit from the supercomputer and in fulfilling the national mission of these centers.

We are currently using 56 kilobit/second communications, using both terrestial and satellite links. When funding for higher bandwidths becomes available, the capability of the system to handle very large volumes of data, such as those obtained in astrophysical observations, and to provide interactive high-resolution graphics capabilities to the users, will be greatly enhanced.

There are more than 200 existing engineering and scientific applications program packages that run on the Cray XMP/48 supercomputer in use at SDSC. But there are two key systems that contribute especially to optimal use of the supercomputer resource. They are the Cray Time-Sharing System (CTSS) and the Central File System (CFS).

CTSS, developed by the National Magnetic Fusion Energy staff at Lawrence Livermore National Laboratory, gives the system its broad range of interactive capabilities and allows hundreds of scientists, some many thousands of miles away from the center, to have direct access to the computer as easily as if it were sitting right next door. That means no waiting for batch results that delay the research effort.

The Central File System, developed by the computing staff at the Los Alamos National Laboratory, is a versatile file storage system for permanent hierarchical storage of large quantities of data and programs. What this means to the working scientist is that the most active files are immediately available; less active files—and files containing very large volumes of data—are located in a second storage tier, while the data from completed experiments is archived for permanent, inactive storage. CFS will automatically migrate files between the three levels of storage as necessary to provide fast file retrieval and maintain optimal use of storage resources.

These features of our system, when added to the utility of SDSCNET, and the links our system has to MFENET and other existing supercomputer facilities and networks such as ARPANET, CSNET, and the emerging NSF Internet, expand the boundaries of scientific investigations that can be carried on through SDSC.

SDSC PHILOSOPHY

Our center is based on a very conservative philosophy: All of the hardware, software, and networking systems are in existence today, and have logged thousands of man years of use. They are not experimental. We know from our own past experience and the experience of others that they mesh together in the most effective computational-communications system available today. The technologies that have been developed through government-funded programs, primarily the Department of Energy, can now be transferred to the broader national research community through SDSC.

A major mission of the San Diego center is to educate a large population of students and faculty in the effective use of supercomputers. So, in addition to making the most powerful hardware and sophisticated software available to users, the staff of SDSC will provide comprehensive instruction in the use of these resources. As this cadre of trained scientists, with hands-on supercomputer experience, moves from academia to industry, we believe a ripple effect will occur, resulting in rapid advances in technology and product development in the United States.

ADVANCES THAT CAN BE EXPECTED

The research done at SDSC will impact all of us, in some areas that are now predictable, and in other areas that in the beauty of scientific inquiry are unpredictable but no less certain, and will be revealed in time.

The kinds of benefits we can now predict include the health and safety effects of better weather prediction, new medicines to treat illness, improved aircraft design, safer automobiles, and increased global food production. The San Diego consortium includes some of the world's leading biomedical research institutions—the Agouron Institute, the Research Institute of Scripps Clinic, and the Salk Institute for Biological Studies—and the medical schools at Stanford University and at the University of California, at Los Angeles, San Diego, and San Francisco; and the universities of Hawaii, Maryland, Michigan, Utah, Washington, and Wisconsin. The consortium can be expected to produce particularly good results in molecular biology, genetic design, and pharmaceutical development. Of course, the range of research will extend to every discipline: astronomy, physics, chemistry, engineering, oceanography, and others.

Research being done today will profoundly affect our lives in the near future. Examples of work by SDSC consortium members currently under way in molecular biology that has been made possible by access to supercomputers include these:

- Robert Fletterick at the University of California at San Francisco has achieved the first modification of a naturally occurring enzyme (trypsin) and has unraveled the structure of the largest asymmetric biological molecule (glycogen phosphorylase). This enzyme regulates blood sugar, and knowledge of its structure has major medical implications.
- Arthur Olsen at the Research Institute of Scripps Clinic has determined the structure of the largest symmetric biological molecule.
- John Ribiere and Arnold Hagler at the Salk Institute and the Agouron Institute have made the first computer model of potential birth-control drugs (peptide analogs).
- Joseph Kraut and Arnold Hagler and the University of California at San Diego and the Agouron Institute made the first use of supercomputers to perform molecular modeling of supercomplex protein molecular system (DHFR methotrexate complex).

Since all of life's processes are based on biological chemical reactions, reactions that are stimulated and enhanced by biological enzymes, the ability to redesign—or design completely new—biological molecules allows us to control those chemical reactions and shape them to the benefit of humankind. Supercomputers are speeding this research and opening areas of study that were not practical in the past.

Studies are also under way to determine the structure of DNA-binding enzymes that will permit genetic alteration of cancer cells, leading to improved cancer treatment, and probably an eventual cure. The anatomic structure of viruses is being investigated—with potential benefits ranging from cure of the common cold and influenza to the cure of more esoteric virus-induced maladies. Or, better yet, to development of vaccines to prevent them.

The ability to model and analyze complex physical structures is the key to major breakthroughs in medicine and many other fields. Our scientists, aided by supercomputer capabilities, can accelerate the process, turning theoretical applications of science into practical, everyday benefits for all of us.

SUMMARY

The San Diego Supercomputer Center demonstrates a unique arrangement. It blends the rich research environment provided by the University of California at San Diego with the experience of GA Technologies in high-technology research and development, operation, and use of state-of-the-art computing systems, as well as practical management skills, bringing together the best of the public and private sectors. I believe this will be reflected in the scientific accomplishments that will come out of our center.

In designing the center, we have adopted a conservative philosophy, using existing systems that have been proven through years of use. No resources will be expended on experimental systems or trying to develop new ones. We will leave that to others.

Interactive access to the computer, combined with the high-speed communications features of the San Diego Center, will accelerate experimental results, producing immediate, practical benefits from the research of hundreds of scientists linked together by SDSCnet and other existing programs networks. In particular, there will be explosive progress in the field of molecular biology, leading to greatly improved medical treatments and even the elimination of some diseases and genetic disorders. These scientists, with the assistance of supercomputers, will also increase world food production through the design of plants that can produce more and are impervious to most plant diseases. They may even develop new crops that can thrive under very adverse and primitive conditions.

13

THE COLLABORATIVE VENTURE AT FLORIDA STATE UNIVERSITY

ROBERT JOHNSON and JOSEPH LANNUTTI

The Supercomputer Computations Research Institute (SCRI) is a four-way cooperative program of Florida State University, the State of Florida, the U.S. Department of Energy, and the Control Data/ETA Systems Corporation. This collaboration between university, state, federal, and industry resources is an excellent example of the emerging concept of "technology venturing."

The following data illustrate how such venturing is structured. The total first-year budget of the Institute is approximately $10 million. Florida State University is providing about 15 percent of the support through funding of computing center personnel and the CYBER 760 front end. The state has provided about 10 percent through funding 13 faculty positions and support positions. The DOE provides 65 percent of the support, of which a major fraction is used to cover the cost of access to the computer for DOE-sponsored programs nationwide. The remainder is allocated to support research by the Institute personnel in computational science. Control Data has provided 10 percent of the funding to support a team of systems and application software specialists on CYBER 205 and ETA–10 computers.

SCRI is recruiting scientists, specialists, and the support staff. In addition, SCRI-connected faculty members are beginning to serve a catalyst role in stimulating interest among Florida State researchers. Supercomputer time itself is a commodity already in short supply, as eight groups of off-campus scientists engaged in energy research projects for DOE, along with five groups of Florida State researchers in the energy research area, have already been assigned time by DOE. The off-campus users represent research teams of from 5 to 15 people each, and include 4 high-energy physicists, 5 nuclear physicists, 3 specialists in materials research, 2 chemists, 3 applied mathematicians, and a molecular biophysicist. On campus, DOE users include a high-energy-physics group, and the Geophysical Fluid Dynamics Institute.

The most serious problem facing the Institute in making these computational resources available to faculty members of other universities is the lack of access by the universities to sufficiently high bandwidth communications capability. If we assume that all the good researchers are found only at major universities, we do not only do a disservice to the smaller institutions that also have excellent personnel, but we begin to ensure that eventually such a statement becomes true, to the detriment of our educational system. Good scientists with a need for supercomputer access are found in many locations, many of which could not justify support for a dedicated supercomputer. Yet what a disservice we would render the nation to suggest these scientists should not have supercomputer access.

It has been suggested that every campus in the nation should have a supercomputer, and perhaps some day the technology will evolve to the point where that is economically feasible. It is not feasible today, but communications capability to make the existing centers available to other institutions is feasible now. Low-cost earth stations consisting of an up-link and a down-link are available for a few thousand dollars, and certainly we have the capability to place communications satellites in the heavens.

It is reasonable to examine whether or not this existing technology can be utilized effectively to make supercomputers accessible to qualified scientists and researchers at all institutions, or just the select few. The primary problem is that supercomputer software development today is a methodology for effectively utilizing multiple processors. With this problem solved, there will be a quantum leap in effective use of super-computers. It may well be that a researcher at some small university somewhere will be the key to that quantum leap—if he or she can get access to a supercomputer. Communications will be the key to that access.

Many have contributed to establishing the methodology of user-oriented communications networks. The various networks such as AR-PANET, BITNET, CSNET, and MFENET have taught us much. The planning and effort that have gone into SCIENCENET have consolidated many of these ideas, and have developed new ones. The time has come to adapt this accumulated knowledge to meet the high-bandwidth requirements of supercomputer technology. Consider what it would mean to have available a 50-to 60-megabit communications path linking the nation's universities. Such a link would not only have the capability of carrying large data sets for supercomputer use but would also serve to carry telephone conversations, television signals, electronic mail, databases for small computers, and in general to serve as a pipeline for linking these centers of learning. It is well accepted that such interaction between our scientists and researchers provides synergistic benefits far beyond what might be expected from individual, isolated research.

For example, there exists one institution that has a very scarce resource in the form of a scanning device that can provide data to assist in the diagnosis of medical problems. The analysis of the data, however, is very complicated. Another institution has the leading expert in interpreting the data. The present practice is to collect the data and send it by mail to the other institution, where in due course it is analyzed, and the results returned. However, with a high-speed communications link, it would be possible to send the data immediately, and to have the results returned by an electronic medium. Many other examples could be cited by our scientists and researchers, covering many different and varied areas of expertise. The sum of the existing needs, plus the new needs that would be generated as experience is gained in using such a system, would pay dividends far in excess of the cost of such a system in terms of knowledge, of efficiency, and in usage of shared resources.

Supercomputers are but one example of resources that can be shared effectively. Other examples can be cited in laboratory instrumentation, in medical tools, and in libraries, to name but a few. The concept of scattered resources that can be shared suggests a communications access medium described previously. The technology exists, in the form of Time Division Multiple Access (TDMA), which allows many earth stations to share a single satellite transponder. Several TDMA systems have already been put into operational service, and more systems are planned. Unfortunately, these systems are not generally available to universities, and it is not feasible for a single university to bear the cost of such a system. With leadership in the concept of shared government-industry efforts, such an endeavor would be a proper effort for government to endorse, and for industry to support.

Florida State has been working with our scientific colleagues within the Southeastern Universities Research Association (SURA) to put together such a plan. SURA would be ideal for a pilot project to demonstrate the feasibility of such an endeavor. We are convinced that unless government is willing to take the lead, no one university or group of universities is going to be able to shoulder the burden.

As the Lax Committee pointed out, the shortage of skilled supercomputer professionals is of critical concern to our nation. Florida State will serve as a primary training center for such professionals as an integral part of the institute's activities. The institute recommends close collaboration with government programs in this area, and that joint programs be offered for training scientists on supercomputer technology. NSF is making such a program available for summer institutes to use the supercomputer facilities they have sponsored. These efforts should be extended by other agencies, such as DOE. Florida State is prepared to assist the NSF or any other agency in implementing such a training program utilizing the supercomputer facilities at the institute.

Examples of activities begun during the first year of operations at the institute are detailed in Figures 13.1 and 13.2.

For the institute staff proper, recruiting has gone forward in a wide range of disciplines. When fully staffed, the institute will have 53 fellows and support staff.

Examples of planned software projects for the institute include a CYBER time-sharing system (like the CTSS system at Livermore), research of vectorization techniques, vectorizer compiler development, techniques and tools for multiprocessing, vectorized and parallelized numerical methods and algorithms, fast partial differential equation solvers, and the establishment of DOE user libraries for the supercomputers we have.

Examples of research projects for which we have employed scientists are shown in Figure 13.1 and include: lattice gauge theory, data processing for high-energy experimental particle physics, accelerator design, many body problems, Monte Carlo techniques, event simulation, meteorology, geology, oceanography, computational fluid dynamics, symbolic manipulation programs, materials by design, molecular dynamics, computer systems, and architectural enhancements.

Figure 13.2 shows the work breakdown for the management responsibilities. The users and advisory committees are currently being arranged.

Recognizing the institute's major support, the program will emphasize research in the computational needs of the Office of Energy Research of the Department of Energy.

In general, the institute's mode of operation in doing research on techniques for large-scale computations will be through their use on actual problems in traditional scientific research. SCRI brings together participants from various disciplines in order to share knowledge regarding solutions to common computational problems.

To promote these goals, the institute is arranging various forums for discussion, including periodic workshops and conferences. Examples of programs already in progress or completed include periodic workshops on CYBER 205, seminars in a variety of fields, a conference on lattice-gauge theory, and a summer school on Monte Carlo techniques.

The institute has two groups of users (a "35 percent group" and a "65 percent group"). The 35 percent group consists of faculty associates and staff at Florida State University and other universities in Florida, who will be allocated 35 percent of the computer access time—via a local steering committee.

The 65 percent group consists of users from universities throughout

Figure 13.1
SCRI Research Through Applications

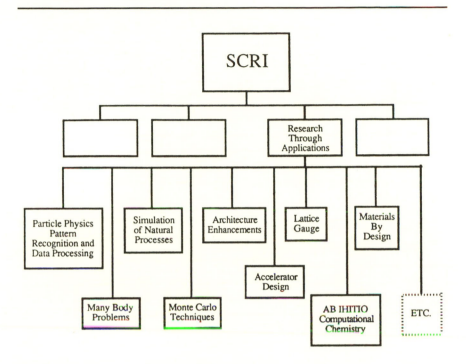

Source: The figures for this chapter were compiled by the authors.

the United States who have been allocated the remaining 65 percent of access time by the Department of Energy's Washington office. At the present time, about 18 of these DOE users have already been allocated all of this portion of the 205 time for the fiscal year. Examples of work being done in various areas by currently approved users of the Florida State CYBER 205 include:

- Applied math: development of algorithms, languages, and architectures for future and more powerful computers
- Materials science: tailoring of materials to satisfy defined requirements
- Biology: DNA and cancer research
- Meteorology: study of time-dependent fluid flows in irregular domains—transition to turbulence, chaotic behavior—in three dimensions
- Mechanical engineering: study of complex, multicomponent computational fluid dynamics

Figure 13.2
SCRI Management

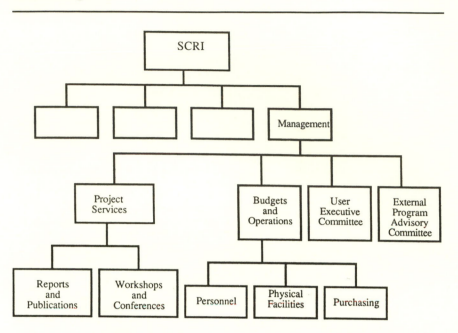

- Oceanography: development of model of Indian Ocean needed to understand and predict interplay of oceans and CO_2

- Accelerator Design: study of highly nonlinear, chaotic effects of beams of particles in accelerators

- Chemistry: study of molecular and atomic scattering

These users, some of them DOE users, are distributed from California to New England, initially are communicating via TYMNET or commercial long-distance lines. However, where possible, they are now being shifted to communication via the National Magnetic Fusion Energy Network.

As to the future, the Institute expects to (1) continue to increase the power and storage capacity of computer facilities; (2) allow and assist increased research utilizing the computers in all physical sciences; and (3) encourage utilization through modeling, expert systems, and artificial intelligence in areas such as music, economics, health care, philosophy, and sociology.

In conclusion, we contend that the SCRI program is an idea whose time has come. The institute is functioning well and serving government, academic, and industrial scientists across the nation. We commend the arrangement to other technological centers and would be pleased to share the experiences of establishing and administering the institute with colleagues.

14

THE INDEPENDENT CENTER AT THE UNIVERSITY OF GEORGIA

WALTER McRAE

The University of Georgia's supercomputer center is not one of the NSF-funded centers. The CYBER 205 (a vector processor) was installed in 1984. A second high-performance computer, the CYBERPLUS (based on a parallel architecture), is already on the machine-room floor, and operational services on this machine began during the summer of 1985. Both installations are combined state-university–industry ventures, and they received no federal funding for start-up costs.

The University of Georgia has always had ability to move basic research from the insulated realm of the laboratory into the practical, and prosaic, realm of real-world applications. Technology transfer is its forte, plus the pragmatic and practical approach being taken to high-performance computing systems and the effective use of today's supercomputers for real-world problem-solving. It is from this perspective that two key issues related to the supercomputer research agenda will be addressed. The first issue concerns how the applications (problems) to be supported with high-performance computing resources are selected. The second issue deals with which activities are supported and how those decisions affect the infrastructure of high-performance computing in this country.

BACKGROUND FOR RESEARCH ISSUES

It is helpful to have a feel for the working environment of the Georgia supercomputer center. Our applications and research projects encompass a wide range of disciplines and fields and are drawn from both the public and the private sectors.

Public-Sector Applications and Research: Environmental Protection Agency researchers are using the supercomputer center to develop national purity standards for bodies of water subject to chemical and other types of pollution. National Park Service researchers are using the computer to analyze the distribution patterns of rare and endangered plant species in the southern Appalachians. U.S. Forest Service fire researchers use the center to model and report weather conditions that are optimal for controlled burnings.

Industrial-Sector Applications and Research: There are applications in digital scene generation for storage on optical disks, the structural design of antennae, and the simulation of process control in composite materials manufacturing conducted at the center.

Academic Applications and Research: The largest single group of individual users of the center are researchers from several universities throughout the Southeast—Kentucky, Tennessee, Alabama, and Georgia. These academic applications range from basic research and engineering to what could be considered production support. Basic research studies being supported include nucleotide sequence analysis in molecular genetics, ecosystem modeling using remote sensing data, beef cattle breed improvement studies, legislative roll-call-vote analysis, adjustable rate mortgage pricing studies, plus more traditional investigations in physics and chemistry.

Applied applications are coming from the fields of chemical, mechanical, and aeronautical engineering. A researcher at the University of Kentucky is investigating aerosol growth and dispersion in the atmosphere, for example. At the University of Tennessee, a mechanical engineering professor is using the CYBER 205 for studies in computational fluid dynamics. Aerospace engineering researchers at the Georgia Institute of Technology are calculating unsteady viscous flows over helicopter bodies and blades, investigating crack growth in solids, and performing buckling analysis on large space structures such as those used by NASA.

To open up the use of the CYBER 205 for an even larger range of applications and user communities, several large software packages that can benefit from the vector processor have been converted and installed, including a number of mathematical and engineering packages distributed by the COSMIC facility. A similar strategy will be used on the CYBERPLUS. For example, researchers in the Georgia center will be adapting PROLOG, one of the principal languages used in artificial intelligence research and expert systems applications, to take advantage of the parallel architecture of the CYBERPLUS machine.

RESEARCH ISSUE 1

The first research agenda issue concerns the criteria and procedures used to select the applications of problems that warrant the use of high-performance computers. Several years ago, when public attention was drawn to the need for national initiatives by groups such as the Lax Committee and the Press Commission, several reports were produced and distributed widely that described a number of areas for which high-performance computing was viewed as essential. There were basically a half dozen areas listed: computational physics and chemistry, weapons research and development, atmospheric and ocean modeling, aeronautical structural research and design, petroleum engineering, and computational fluid dynamics. The authors of these seminal reports undoubtedly intended these lists only as examples, but, unfortunately, the lists are being picked up and used as if they were a definition of the problem domains for which high-performance computing is justified. Even worse, they are often interpreted as defining the only types of applications for which supercomputers are appropriate.

As is apparent from the uses already being made of the Georgia center, this is a faulty perception and much too narrow a view of the opportunities. There are many problems and applications that did not happen to be cited as examples in those original reports that can benefit greatly from being moved to these machines.

If it were only a matter of misperception, it wouldn't make much difference. Unfortunately, those lists are increasingly being used to guide policy and funding decisions—in effect, setting a de facto national agenda of what constitutes valid problem domains for supercomputer research and use. Those original lists are also influencing support decisions in agencies whose primary missions lie outside the specified areas. Scientists and engineers with valid needs for access to supercomputers are finding their requests for computing support turned down because their application may not be one of these six illustrative application areas. The selection of supportable problems is also being influenced, and inappropriately so, by some misconceptions that if the proposed work in any way uses high-performance computers, then it should be funded (or at least approved) by one of the agencies directly supporting research initiatives in this area—the National Science Foundation or the Defense Advanced Research Program Agency, for example. A distinction needs to be made between research about high-performance computers and use of such machines. While some projects do, indeed, involve both research and use, many others don't, and our national resources will remain seriously underutilized if these misconceptions are allowed to bias problem selection.

The requirement that NSF funds for supercomputer use be spent only at one of the NSF-funded centers also constrains the selection of problems and could have a chilling effect on use of these machines, as well. Because of this policy, researchers from our institutions who are awarded NSF grants that include supercomputer time must either forego the NSF funding support or must take their work off the machines located on their own campus or attached to the statewide network and move it to one of the out-of-state NSF-funded centers. Not only is that counter-productive and inefficient for those researchers, it also runs counter to prevailing free-enterprise philosophy. In the business world, such a policy would probably be tantamount to restraint of trade.

In setting the research agenda, care must be taken so that choices and problems are not arbitrarily biased or constrained, either intentionally or inadvertently, to some small number of highly specialized target areas. To allow this to happen is to run the risk of missing many valuable opportunities to exploit a valuable national resource. Furthermore, a narrow or highly specialized focus leaves the country vulnerable to com-petition from abroad in those areas not supported, and hence weakens the overall fabric of the nation's research and technology framework. Large research universities are in a unique position to ensure that the broadest range of national needs are given due consideration and that all users, regardless of disciplines or applications, are provided equitable access to these computing resources provided perceptions or policies do not arbitrarily constrain the valid problem domains.

RESEARCH ISSUE 2

The second agenda issue to be addressed concerns the types of work supported, particularly the balance of support across the research-de-velopment-use continuum, and the implications of those decisions for the infrastructure of high-performance computing in this country.

In addressing the nation's needs for next-generation computing, at-tention to date has been focused very heavily on basic research, espe-cially basic hardware research. It is hard to pick up a journal or trade magazine that does not contain articles on nanosecond memory access times, submicron VLSI architectures, picosecond logic gate times, ter-abyte storage devices, gigabit data transmission rates, and megabit data communications rates. This preoccupation with computer hardware is understandable, given the nation's desire to lead in the next generation of computing machines. However, hardware alone is not enough to insure that leadership position. The fastest processor is nothing more than an electronic curiosity if it lacks the software to turn it into a useful tool. Attention must also be given to other aspects of the computing environment if the nation is to have a solid infrastructure that fits to-gether without gaping holes.

If the present is any predictor of the future, the full range of needs is not yet being adequately addressed. The contribution to the infrastruc-ture illustrates some of the deficiencies well.

Even the best of the current supercomputers—the Crays and the CY-BERs—have software that is little better than what was available on the large scientific computers of 20 to 30 years ago. The operating systems and compilers are primitive; the editors and file management utilities, which come standard on state-of the-art mainframes and minis, are virtually nonexistent. For researchers to move their work from their campus-based mainframe computers to one of the high-performance machines is, in many respects, a giant step backward in data processing techniques. This vacuum in software also induces training and learning requirements that would be unnecessary, or at least minimized, if appropriate software tools were in place.

Although this vacuum in software support has been recognized and is, to some extent, being addressed through initiative based mostly on work-station design, sufficient attention has not yet been given to this critical area. More important, there is little indication that sufficient attention is being given to coming up with an *integrated* solution that adequately takes into account not only the state of the operating environment for the supercomputers themselves but also the most appropriate distribution of functions among individual researchers' work stations, mainframe front ends, and, ultimately, the high-performance engines. The configuration of the communications network that provides the access highways to these facilities should be greatly influenced by how these functions are distributed. While work is going on on some of these pieces, it is largely a piecemeal, uncoordinated, unplanned effort.

As an item for the research agenda, an activity—an identifiable project, perhaps—is needed to lay out a modest set of reasonable scenarios for how advanced scientific computing could develop over the next few years. These scenarios then become models against which proposed solutions can be evaluated within a framework where the assumptions are at least explicit and the likely outcomes apparent. Until such scenarios (models) exist as a general framework for decision making, conflicting and often counterproductive courses of action will continue to be taken, driven in large part by special-interest needs. Our perspective, vis-à-vis high-performance computing and its supporting infrastructure, needs strengthening and broadening to provide a common framework for decision making.

SUMMARY

In summary, two recommendations for the research agenda have been proposed:

1. to broaden the perspective of what constitutes valid applications for use of high-performance machines, taking positive steps where necessary to remove barriers or constraints that have crept in through funding or support policies;

2. to broaden the perspective of R&D activities critical to the overall success of high-performance computing, giving special attention to the interrelation-ships of these activities and their implications for the national infrastructure.

Because of their broad missions and research interests, as well as their pool of talent, universities should be able to play important roles in addressing these two agenda items.

PART V

INDUSTRIAL APPLICATIONS OF SUPERCOMPUTER TECHNOLOGY

INTRODUCTION

Supercomputers are rapidly becoming a major element in the operations of U.S. industry. Confined earlier to scientific and defense research in federal laboratories, supercomputer applications emerged first in the private sector as research tools for the aerospace industry. Spurred by development of space vehicles and advanced airframes, companies such as Lockheed and Boeing led the way in computer-aided design, and developed the concept of three-dimensional modeling as a replacement for wind tunnels.

Following the examples set by aerospace, other U.S. industries have quickly integrated supercomputer technology into their research, design, and development operations. The automotive industry has reduced costs and improved productivity and quality control through imaginative use of supercomputers. The energy and chemical industries have found supercomputers key to geological exploration and substance analysis. Nuclear research in both power generation and the sticky problem of waste disposal has taken a quantum leap forward with the application of supercomputer technology.

As industry has developed new applications for supercomputers, manufacturing firms have become integral members of this new community and are seeking a significant role in policies and programs impacting the future of supercomputer technology in the United States. Chief among their concerns is the need for scientists and engineers trained on supercomputers. Without adequate, specially trained scientists familiar with the new machines and their capabilities, industry cannot make full use of supercomputers and cannot realize maximum return on the investment in hardware and related peripheral equipment. A recurring theme in industry is the need to establish effective communications networks between companies, universities, and government-funded supercomputer centers. Access to scientists and their research capabilities utilizing supercomputers is a key element in increasing the productivity of private-sector research and development.

The chapters in this section represent a cross section of industrial users of supercomputers in the United States. They are indicative of how U.S. companies justified acquisition of supercomputers, how they have benefited operations, and their recommendations for initiatives to enhance utilization of these new tools.

15

THE AUTOMOTIVE INDUSTRY
PRODUCTIVITY, COST, AND QUALITY: SUPERCOMPUTERS AT GENERAL MOTORS

GEORGE D. DODD

In the fall of 1983, General Motors Research Lab added to the research computing a Cray-1S supercomputer. This system is being used for research rather than application or production computing within General Motors (GM). The Cray was added to a family of already existing computers—IBM 3081 and VAX—tied by means of local area networks and direct ties to the terminals and personal computers throughout the laboratories and throughout the corporation.

The role of supercomputers in GM's industrial research laboratory can be seen by describing a typical research application using the supercomputer, and discussing the supercomputer's impact on our research. Our experience resulted in some surprises (mostly positive) and points toward future supercomputer requirements.

TYPICAL RESEARCH APPLICATIONS

Aerodynamic research provides a useful case example. As we begin to look at model aerodynamics, we first have to develop a model of both a vehicle and the environment. The model of a 1984 Pontiac Fiero consists of about 3,000 panels. The data for this come from a clay model processed by a General Motors computer design system called CGS, rather than by means of networks into research laboratories. A second model of the air field surrounding the vehicle also has to be developed.

The number of elements modeled in an aerodynamic model varies widely. In the nonviscous flow region, which is the area removed or remote from the vehicle, there are about 20,000 to 30,000 elements where friction of the air as well as friction against the vehicle surface becomes important. The three-dimensional models are upwards of 100,000 node points. The model size being maintained is roughly 5 million eight-byte words, or 40 million bytes of memory. This has presented a significant problem in the solution of aerodynamic models because of the lack of large memories in which to solve the situation. We can either go to the out-of-memory solution process, which dramatically increases the computation time, or reduce the model size and therefore reduce the designed accuracy we are trying to achieve.

Several equations are solved simultaneously regarding the maximum and minimum turbulence of the air surrounding the vehicle. All six to nine equations per element or per panel are solved within the final modeling process. Once the equations have been solved, we need to extract the answers.

For an aerodynamics model there is a fairly small amount of data; roughly 600 data points need to be evaluated. This is about 1,000 pages of paper, which is conceivable to compute manually. In modeling engine combustion systems, which is a time-varying situation, there are results with 3 to 4 billion data points of information that need to be evaluated. Clearly, this is far too much information to compute manually. The only way to do this is have the computer system help extract the information and also help plot the graphic results.

This is a highly iterative process consisting of developing the model, computing the solutions, and evaluating the results. Initial results do not work as expected. Consequently, it is necessary to change the model, recompute, reevaluate, and so forth. About 40 percent of the total time solving aerodynamic problems is spent in developing the models, about 20 percent in doing the arithmetic, and 40 percent in evaluating the results, generating graphics, and getting ready to go back and think about the process again.

This process requires a tight, high-speed coupling of the entire environment. It involves the modeling, which takes place perhaps on one set of computers in one area of the corporation; the computation, which occurs on the supercomputers; and the results evaluation process, which might take place on other computers and require the use of graphics terminals and other graphics facilities. This tight coupling is vital to the timely analysis and solution of problems run on GM's supercomputers.

IMPACT OF THE SUPERCOMPUTER

The primary impact of supercomputing on research at General Motors has been that the supercomputer is used to validate laboratory experimentation by modeling and guiding the experimental process. Before GM's supercomputer was installed, the computers were used primarily to do arithmetic in support of laboratory work. In retrospect, these were fairly trivial calculations. Now the reverse is true. The laboratories are used to verify the modeling of the supercomputer and to allow the user to ask "what if" questions. Why is the model behaving this way? It allows us to help focus the laboratory research to determine the answers to those questions and then feed that back into the modeling through the next iteration.

The second major impact is that GM is now exploring three-dimensional-model solutions. Before the installation of the supercomputer, most of our work was two-dimensional. Now with the 3–D modeling capabilities and the ability to handle much larger environments, we are combining separate models. For example, our research in chemistry of combustion in the cylinder of an engine requires two separate models— one model of the flow of fluids in the combustion chamber in the moving piston, and the second of the chemical combustion kinetics of mathematical determinations. We are now modeling these into one complete model, which requires much more intensive computation, but which provides answers that are much more accurate. We are able to prove the accuracy of the model by having many more grid points in the model and to do a much better job of evaluating issues around areas of great concern.

Last, we are speeding up the pace of experimentation. Before the supercomputer was installed, it might have taken several weeks to get one answer through nighttime running on our largest IBM computers. That same research work is now being done at a pace of several answers per day.

In addition to speeding up experimentation, we are able to run pieces of models to examine critical parts of the entire environment and then use that to help direct laboratory research, which in turn is fed back into the modeling process.

SIGNIFICANT EXPERIENCE GENERALIZATIONS

Moving through the operation of our supercomputer, we encountered a number of surprises. First and foremost, as we anticipated, was its delivery. We thought the speed would be a limiting factor on the use of the machine. It turned out that memory clearly was the limiting factor. Some of our combustion models are upward of 100 million words in size. Clearly, this is way too large for anything possible at the present time.

The FORTRAN conversion was easier than expected. Converting the FORTRAN 77 on the IBM to the Cray worked well because of good diagnostics and a very sophisticated vectorizing compiler. This is a key point in trying to spread supercomputers to the masses. We must have an easy way through an existing code onto the supercomputers because of the range of machines—from small "Crayettes," to the very large machines, to the machines on the desk.

The required operational support was lower than expected. We had anticipated a certain amount of operators in the machine room, as well as programming support. This came in much less than budgeted. What we had thought might be big problems turned out to be not so big. They only whetted our appetite to begin to solve truly big problems that are now being worked on.

Finally, we are beginning to see a movement elsewhere—in the mathematical world, the world of vendors, the universities, and government laboratories—to generate mathematical packages optimized for supercomputers and more easily installed in a supercomputer world.

FUTURE NEEDS

Future supercomputer requirements based on the work we have done to date are as follows: First, we need much larger real memory. I emphasize the word *real*, not virtual, because virtual memory implies an I/0 process that greatly degrades the computational activity. Of course, speed is important. We would always like to have more speed. But memory in our physical environment is much more important than speed at the present time.

Second, we need to find a way to couple supercomputers with large data bases. Traditional data-base technology implies sequential access of records and files, which is good for payroll and inventory control purposes, but not good for large in-memory of calculations. We have a vector-type supercomputer. Research must be completed and products developed dealing with a vectorized network or vectorized data base. Likewise in the area of parallel supercomputers, data bases should be of parallel form.

A significant amount of time was spent not in solving the problem but in evaluating the results. Methods need to be developed to relate the graphics to the input models or generate the graphics as a result of the output and feed that back into the input process.

Last, we need to explore alternate architecture for problems that do not lend themselves well to vectorization. My element models, for example, are models relating to solid geometry such as crash-simulation and image-processing models. These clearly do not do well as vectors. We need to explore all architecture that relates to parallel processors.

There needs to be very-high-speed links to tie the supercomputer back to the other computers within the environment and to the users to generate the input and to get results on a timely basis. This should be a high priority of both government and industrial users of supercomputers.

SUMMARY

In conclusion, the supercomputer at General Motors has had significant impact on our laboratory research. The growth and use has been more than expected. Many opportunities have emerged for use of the machine because of its availability and because it allows people to think about problems they otherwise would not have considered.

THE EMERGING NEED FOR SUPERCOMPUTER-TRAINED ENGINEERS AT FORD

HENRY ZANARDELLI

The once-dominant status of this country's domestic automobile industry has been seriously challenged during the past decade by foreign competition—especially by Japan. Some of the factors behind this assault are beyond our industry's control, but others are not. Our ability to control two key factors in this competition—product quality and product cost—is dependent to a great extent on our effectiveness in using supercomputers and related technologies.

The computer is a significant catalyst in helping almost all organizations within Ford to do more and better work at less cost, whether the work is technical or administrative in nature. My focus will be the use of supercomputers in Ford's product development activities, especially explaining how these computers help us to evaluate the function and the performance of the vehicles we design.

DESIGN EVALUATION

Three Steps

Use of supercomputers for design evaluation usually involves three basic steps:

1. Development of a mathematical representation or structural model of the part, assembly, system, or vehicle being designed. This requires the skills of a specially trained engineer.

2. Processing the model through computer simulation and analysis programs. Supercomputers are needed in order to get this job done in reasonable time.

3. Interpretation of the results by an engineer. At this point the achievement of design objectives is verified or needed design changes are identified.

Use of Finite Element Analysis in Models

The introduction of supercomputers along with the desire for faster, more cost-efficient processing of alternative design studies has led to the construction of larger and larger mathematical models developed through the use of a technique called finite element analysis, often abbreviated FEA.

Ford is among the earliest users of FEA. The specific FEA technique we use is based primarily on software originally developed by NASA and called NASA Structural Analysis, more commonly known as NAS-TRAN. (This is an outstanding example of a government space program by-product that benefits the private sector.)

Ford uses FEA with good results in evaluating the design of individual components as well as large assemblies such as engines, transmissions, suspension systems, and complete body structures. The basic concept underlying FEA is that the design of every component or structure can be represented in a model consisting of a series of individual but inter-related finite elements. Typically, a large finite element model will involve from 15,000 to 20,000 discrete elements, with the physical property of each element defined in the model. Then, by means of a supercomputer, trillions of calculations are performed to determine how the design will behave under simulated loads and operating conditions.

A supercomputer capable of hundreds of millions of instructions per second is needed in order to get the job done in a reasonable time. The results of this process are predictions of the deformation, stress, and other physical responses that the component, assembly, system, or vehicle being designed is likely to undergo in actual on-the-road situations.

FLEXIBLE DESIGN ALTERNATIVES

In addition to the use of supercomputers and FEA for structural analysis, we apply other computer technology extensively in our efforts to improve vehicle handling, ride, braking, fuel economy, aerodynamic efficiency, emissions, noise, and other factors affecting vehicle performance.

The use of supercomputers gives our engineers more analytical insight and time to improve their work. Several design alternatives can be explored before selecting one from which a prototype will be fabricated and tested. As a result, we reduce the number of prototypes built, we shorten the time to develop new vehicles, we improve product quality, and we save on design and test costs. As one illustration of this, our experience has shown that, when feasible, computer simulation of a design test sometimes can be done at one-fifteenth the cost of a comparable test using an actual prototype.

A LOOK TOWARD THE FUTURE

While finite element analysis and supercomputers already play important roles in Ford's product development process, use of these technologies is still in its infancy. We want to model more complex vehicles and systems.

Our Company

We want to continue maturing in our capability to simulate destructive testing of entire vehicles, expanding our present capacity for producing both visual and digital output used to analyze the crashworthiness of some designs. We look forward to the day when more automotive styling can be done by computer, and when we can determine the aerodynamic behavior of car and truck designs without the aid of physical vehicle models and wind tunnels. There is still a long way to go. Even though we already have experienced some major upgrades, more sophisticated software and even faster supercomputers are needed.

Convinced that the desired tool will continue to emerge in a cost-effective way, we anticipate that our engineering work in the future will involve a paperless process—one through which our designers and engineers will develop and test their designs by computer, storing their results on data bases to be used by the downstream manufacturing activities that fabricate the dies and machinery needed to produce parts for our vehicles.

We already employ a long-term training and education program for our engineers so that they gain an appreciation for computer-based product design and acquire the skills needed to apply it. To some degree, this is a cultural change for them, as they were not exposed to the capabilities of modern supercomputers while in college.

Academic Institutions

Use of supercomputers in U.S. industry may be restricted not by hardware or software limitations, but by the limited number of people who have the know-how to use them. Universities and technical colleges need encouragement to adjust their curricula so that there is more instruction in the use of supercomputer technology. More students have to be graduated with an understanding of the supercomputer.

Supercomputers (and the facilities, software, and support personnel they require) are expensive. The acquisition and installation cost can easily exceed $10 million and few schools can afford them. Those schools having a supercomputer often tie it up with scientific research projects and only a handful of graduate students gain exposure. A means for resource and cost sharing by smaller schools is needed, possibly achieved through the use of supercomputer networks like those provided by some computer service companies.

Government

While the government is to be congratulated for its recent initiatives to foster supercomputer research at a few large, prestigious schools, more is needed in the way of aid to the smaller institutions who together produce the bulk of our engineering graduates. Much more needs to be done to assure our nation of a continuing and adequate stream of up-to-date engineering talent, and to facilitate the needed programs and incentives.

SUMMARY

Ford Motor Company is a leader in the industrial use of supercomputers; it has invested heavily in computers, related facilities, software development, and training. Ford will continue to do so in order to produce safe and attractive high-quality products, at a competitive cost, that customers will want. The vitality of this nation is becoming more and more dependent on the ability of its people to continue to develop and use advanced technical tools such as supercomputers.

16

THE ENERGY INDUSTRY: SUPERCOMPUTER TECHNOLOGY AT ARCO

EDWIN B. NEITZEL

It is not generally known that the largest two users of supercomputers are the U.S. government and the petroleum industry. Both large-scale reservoir simulation and seismic exploration create the need for supercomputers by the petroleum industry. Currently, large supercomputers are highly vector oriented. Most of these vector operations were originally conceived in hardware during the early 1960s as a requirement for economic processing of the multichannel serial time functions comprising seismic needs in the past. It is of interest to examine future seismic needs and hypothesize how these needs might dictate the next-generation supercomputer.

Special characteristics of seismic data processing require the use of supercomputers. Let us consider how the magnitude of seismic effort in the free world is tabulated, how data are acquired and processed, how a typical computer system is implemented, and how seismic needs dictate the required speed of the next generation of supercomputers.

The Society of Exploration Geophysicists (SEG) maintains statistics on the use of the seismic method. The latest report, for 1983, shows that there were slightly less than 500,000 miles of linear land data acquisition at a cost of about $4,000 per mile for a total annual free-world land market of $2 billion. For marine acquisition, there are slightly fewer than 1 million miles per year at a much lower cost ($600/mile) due to the automotion of the data-acquisition ship. For data processing, we have the amount of input data at roughly 200,000 miles at a cost of $700 to $800 per mile for a total market of about $1 billion. In summary, the total market associated with seismic exploration has been varying from about $4 to $6 billion per year for the free world over the last 20 years.

Seismic business is cyclic, as shown in Figure 16.1, as thousands of linear miles versus calendar time for marine and land data. The cyclic

Figure 16.1
Worldwide Seismic Activity in Line Miles, 1964–1983

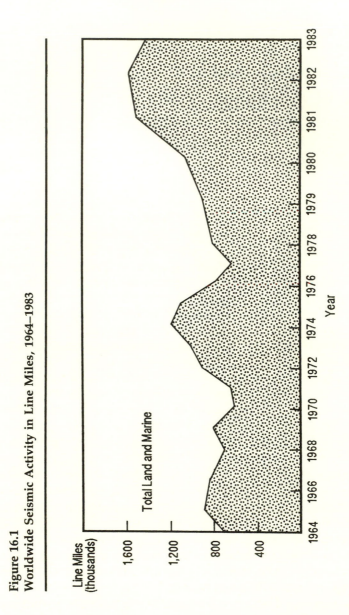

Source: Reprinted by permission from *Leading Edge* 3, no. 7 (July, 1984), 54.

Figure 16.2
Prestack Depth Migration, 5–45 Hz

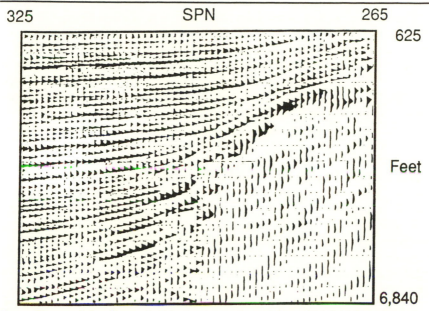

Source: The remaining figures in this chapter are from Atlantic Richfield Company.

nature of the market is shown, but the overall trend through the years has been up. In 1984 the market was relatively flat, and in 1985 it dropped significantly due to the price trends of petroleum. As the price of petroleum increases in the future, this overall trend will come back to an upswing, dictated by market conditions.

FUNDAMENTALS OF THE SEISMIC METHOD

The seismic method is basically an echo method. A signal is initiated, either on or near the surface of the ground, and the acoustic energy propagates down to a reflecting interface. The reflecting interface is a change in acoustic impedance (the product of velocity and density), and the reflected event comes back to velocity geophones on the surface. Large numbers of source initiations and receiver groups are used for actual recordings. A cross section of two-way travel time or depth versus surface traverse distance is then created.

An illustration of a modern record section, which is a cross section looking horizontally into the earth, is presented in Figure 16.2. These data are for a Gulf of Mexico piercement salt dome illustrating the various sediments truncating against the side of the dome.

DATA ACQUISITION

For land acquisition, the seismic impulse is initiated on or near the surface of the ground. Vibrator vehicles, which operate synchronously, move down a dirt road. Adjacent to the road are a series of velocity detectors (geophones), which are placed on the surface. Normally a vibrator impacts an oscillatory signal on the ground surface, which is like a chirp or FM signal, for about 15 seconds and then waits for approximately five seconds for a total of 20 seconds. Then the vehicles move for about 20 seconds to the next location for recording. Therefore the time required for taking a record on land in an automated fashion by a vibrator crew is on the order of 40 seconds to one minute.

A marine survey vessel conducts similar processes offshore. The ship takes data at a continuous linear speed of about four to six knots. Air guns, which release 2,000 psi of compressed air, are normally used for the seismic source. The reflection detectors are in a streamer cable. The streamer cable is normally from one to three miles long and is towed in the water behind the boat at a depth of about 30 feet. The ship *ARCO Resolution* records 240 channels of data. A record is taken every 25 meters along the subsurface. The data are telemetried to the boat where it is properly formatted before forwarding to the large computer center.

DATA PROCESSING

Seismic data processing is performed to mathematically remove many of the distortions caused by wave propagation to and from the reflector and to improve the signal-to-noise ratio. The state of the art in data processing is illustrated in Figure 16.3 by a series of cascaded learning curves. Acoustic modeling basically means that we mathematically comprehend the earth as a fluid for simplification but actually the earth is a solid. By being a solid or elastic model, the earth supports both compressional and shear wave signals, and allows mode-converted signals at impedance interfaces.

When the full elastic model is comprehended in the future, there will be the potential to extract rock and fluid properties reliably, and to define a more accurate structural picture.

After the data are processed they move to other computers for interpretation. The smaller computer connects to various graphic work stations for interaction with the interpreter. Data can be brought up to the screen of a CRT, the interpreter can interact with the data and reach a decision relative to his or her final drilling recommendation. Figure 16.4 provides a side view of an area that was previously mapped. Fault interpretations and many different views of the sides or interior of the data volume can be displayed.

Figure 16.3
Seismic Technological Progress

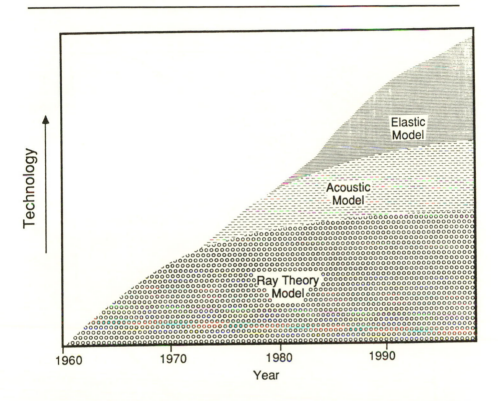

The directions that the industry is addressing in research for the interpretation of data are illustrated in Figure 16.5. The data are organized into structured data bases. We work with the data through interactive graphics to help the interpreter reach a final decision. When we are able to place the logical reasoning power of the interpreter into computer algorithms, then we have artificial intelligence that can result in a step function change in the overall effectiveness and efficiency of exploration.

COMPUTER SYSTEMS

Characteristics of the computer systems are dictated by a need for large data storage. Each one of the input records that is taken each few seconds on a seismic crew has over 200,000 words of information. Up to 100 records per mile are taken in the field. We have to get these records (long serial time functions) into and out of the computer, so the processing is very input-output intensive.

Figure 16.4
Side View of an Area Previously Mapped

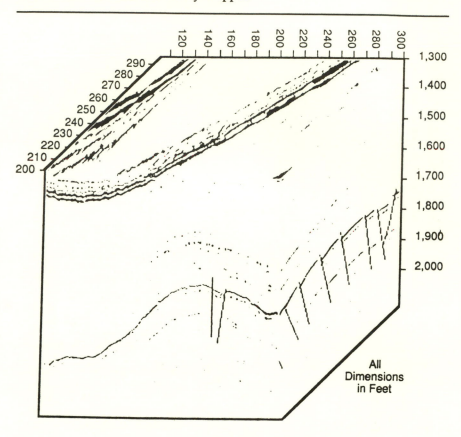

To give you an illustration, Figure 16.6 shows the Atlantic Richfield computer complex. Two large IBM computers interface to various devices. For extra computational power, there are eight array processors and the Cray XMP. To get the data in and out of the computer, over 100 tape transports are employed. Over 300 terminals are used by the people who control the data processing.

The types of operations performed are either vector operations on long serial time functions or scalar operations. Computation times for two simple operations are shown in Figure 16.7. A 9–point operator is convolved across 1,000 data points. The computer times (using only scalar operations and without an array processor) are compared to the times required by a totally vector-oriented machine. The result is a 30:1 ratio to as much as several hundred to one speed-up with the vector machine.

Figure 16.5
Technology Directions Data Interpretation

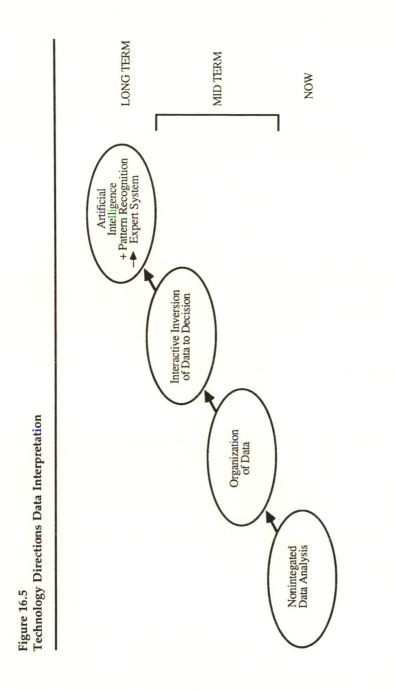

Figure 16.6
ARCO Seismic Computer System

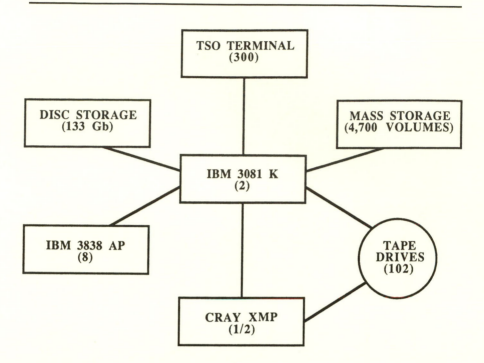

Figure 16.7
Vector versus Scalar Time Comparison

	CRAY - XP	IBM 3081 K	SPEED RATIO
REAL FFT 1024 POINT	1.84 MS	56.6 MS	30:1
CONVOLUTION 1029 X 9	0.12 MS	20.2 MS	160:1

Today, to process a 3–D depth-imaged data volume taken over ten square miles would require about 1,000 hours on the IBM 3081 or 60 hours on the Cray XMP. Three-dimensional depth imaging is normally used for oil-field extension surveys.

With the Cray computer we can do more complex processing that involves interactive interpretive solutions for modeling studies. We assure a solution and compute the theoretical response and see how the theoretical response compares with the actual field response. To compute the theoretical response for a 2–D acoustic earth that is 7,000 feet long and 2,500 feet deep takes 15 minutes CPU time on the Cray. The more realistic elastic case takes 450 minutes on the Cray, a totally impractical amount of time from an economic standpoint.

In the future we need larger supercomputers. More accurate processing of the ten-square-mile example with full 3–D depth imaging would take 6,000 hours on the Cray XMP. This is impractical. The use of next-generation supercomputers, which are forecast to be 100 times faster, might result in a more reasonable 60 hours for this important petroleum exploration solution.

We conclude that seismic data processing is a significant influence in the field of large-scale scientific computers and is of major value and influence in our free world economy.

17

THE CHEMICAL INDUSTRY: TRAINING SUPERCOMPUTER SCIENTISTS AT DU PONT

DAVID PENSAK

Richard Hamming's comment that the purpose of computing is insight, not numbers can be extended to explain Du Pont's interest in supercomputing. That is, we try to use supercomputer technology to generate super insight, not super numbers.

COMPUTING-RELATED PROBLEMS AT DU PONT

What kinds of problems is Du Pont interested in that have relevance to ultra-large-scale computing? The following are just some examples of why we are interested and what we are doing about it, and what we see as the problems that require active cooperation of the entire scientific community to achieve a reasonable solution.

The first example is the process of developing special, semipermeable membranes. It has been estimated that by the mid–1990s the United States will be producing in excess of 10 billion gallons of alcohol a year by fermentation of biomass (or garbage, if you will). This procedure produces about 6 percent alcohol in a slurry of water, sludge, and many other things. It has been estimated that at today's energy prices it costs in excess of 70 cents per gallon to distill that alcohol away. Accordingly, this constitutes a $7 billion a year opportunity if a company could design a semipermeable membrane like a filter paper on which the sludge is poured and the alcohol comes out the other end. Du Pont scientists tried understanding the basic physics of how to design this membrane and finally estimated that if we had a dedicated Cray XMP, by the year 2010 we could have completed a calculation at just today's level of knowledge. There clearly is a real need for a more timely solution. Advances in supercomputer technology will help develop this and other markets.

Du Pont also has a problem in one of our chemical manufacturing plants where we have a reactor roughly 60 feet long and 20 feet in diameter. It runs at about 800 degrees centigrade. The problem is how to optimize the geometry to improve our yield. After about 20 years of research, there were something like 1,800 simultaneous differential equations that precisely describe the behavior of this system. I don't know the number of hours that it took on an IBM 3081 to solve this problem, but the operational savings just in that one reactor could easily reach seven figures annually.

At another plant, we run a reaction at almost 1,000 degrees centigrade and then quench it down to room temperature in roughly 50 milliseconds. That is indeed a major engineering feat. However, the process produces about one part per million of an extremely toxic side product. The problem is: Should the reaction be cooled down faster or slower in order to eliminate that side product (which is very difficult to dispose of safely). We have had to do some very complex kinetic simulation with this problem. It is not yet completely solved, but work is well under way.

WHY DU PONT NEEDS SUPERCOMPUTERS

One of the questions Du Pont is asked frequently is: Given that so much CPU time is spent on chemical computations today that cannot be validated experimentally, why do we want to do more and faster and bigger? It has been demonstrated that the Schrodinger equation cannot be precisely solved for anything bigger than the hydrogen atom, so what we have to do is make assumptions. The more assumptions one makes, the more questionable the validity. It is necessary therefore to do enough detailed calculations to determine how inaccurate the small calculations are. It is really a choice between the best possible result and the best result possible. The only thing that saves us in this is that it has been demonstrated that the mathematics of Hartree Fock theory is a variational calculus. As such, as long as you get a lower number for the total than you got before, it must be a better result. We have calculations that routinely generate between 500 million and 2 billion two-electron integrals. These have to be processed dozens of times in what is called self-consistent field calculations to compute exactly where the orbitals are. This is quite CPU intensive, so we had been buying time on commercial supercomputers and finally acquired our own.

There are certain classes of problems that no general-purpose supercomputer is going to solve for us. For example, Columbia University and Brookhaven National Labs are developing a special-purpose processor (for themselves and for Du Pont's life sciences effort) specifically for the calculation of molecular dynamics. What we are concerned with is trying to see the forces on each atom from all the other atoms in the molecule as they are interacting. Molecular dynamics is an extremely time-consuming procedure. Because of the need to take such very small time steps, we can simulate less than one microsecond of an enzyme's behavior in an hour and a half of supercomputer time. Since some of the physical processes that we are trying to study are taking literally tenths of seconds, it becomes a prohibitive problem. The Columbia-Brookhaven system is specifically designed for this application and we estimate by the time it is finished, it may run several times the speed of any of the Class VI machines. It is costing significantly less than the purchase of a general-purpose supercomputer.

RESEARCH IN PARALLEL ARCHITECTURE

Du Pont is also conducting research into concurrent or parallel architecture. Why is Du Pont interested in these architectures? Thinking back to the molecular orbital problem, the mathematics of self-consistent field theory are quite complex and not physically intuitive. The problem is we have to understand how the molecules actually do behave. If I were to show you the ground state of a molecule and then inject additional energy to make an excited state and asked you to describe how the orbitals relax to try to accommodate this, you probably couldn't. What you could do is a large molecular orbital calculation that would ultimately produce the final numbers. What you would not have is a mental picture of the basic physical processes that are taking place. We need to do our calculations even faster as our molecular systems get bigger and bigger. We believe that a parallelism is one approach that we can take.

A related problem is how to use graphics and to watch what have traditionally been number-crunching processes. How do you change the scientist's intuition so he or she can generate insight and not just numbers? I believe there is a computational analogy to the Heisenberg uncertainty principle, in that as soon as you try to measure a running program on a computer system, you are perturbing it. We have work in progress trying to figure out how to have a parallel computer watch itself. We don't have any good solutions thus far.

PEOPLE AND COMMUNICATION NEEDED

One of the problems in supercomputing now is that there are just not enough trained people. Du Pont hopes universities will increase the number of students trained in supercomputers. We support the National Science Foundation supercomputer initiative as an avenue to facilitate this.

Another problem is that there is not enough communication among the academic laboratories, government, and industry. The perception

that many of the computer-science and computer-engineering depart-
ments have of industrial computing is woefully out of date. Du Pont
more than once has had a professor come through and say, "I have a
candidate for you. He will be great in industry, but he could never make
it in the university environment. He's just not good enough." After a
while one develops a thick skin, but it is a fundamental problem, the
solution to which is of importance to everyone. The difference between
industry and university is that industry has real-live practical problems
that need to be solved in a relatively timely manner for which there is
financial justification. We often have massive quantities of data—often
far more than the university environment has access to.

We have heard the supercomputer consortia all say, "we want industry to send us money and your best resources and we will enlighten
them." I challenge the consortia and the universities to think of ways
where they can also send their people out into industry to get a better
understanding of the real-world problems that we face. Once we start
working together, a lot of the difficulties will go away. Once the contact
is more effective and more efficient, the funding will flow naturally and
there won't be any question in anyone's mind that it is a good thing to
do. No company right now is big enough to try to do everything entirely
on its own. Du Pont would like to find better ways to work with all the
scientific computing community and funding really is not going to turn
out to be a problem when communication is made more effective.

OTHER ISSUES

Another issue is that there is a proprietary aspect to much of the work
in industrial research. There will be times when industry will have to
do calculations that cannot and will not be published. Certain installations have guidelines absolutely forbidding any proprietary calculations
under any conditions, and this puts industry in a box that there is no
way to work around.

Another problem is career development opportunities. When I try to
hire a computer scientist, I have to be honest about career possibilities
in the biggest chemical company in the world. Even if Du Pont ultimately
hired five computer architects, they would not have the kind of pro-
motional opportunity they would have in a computer company. Mechanisms are needed to provide joint appointments in industry and
universities. Perhaps trading people back and forth between government
labs (as long as proprietary information is not compromised) is a route
to give people a chance to achieve the kinds of careers they want.

The last issue causing us a lot of headaches relates to the fact that
most scientists these days know so much about computing that they
tend to design their algorithms with too much knowledge of the imple-

mentation details on their architecture. If you have two molecules interacting, the physical reality is not simulated by nested DO-loops, yet
that is the way a lot of our people think during code development. We
need to build new software tools so the algorithms can be designed from
the very onset to be, if not architecturally independent, at least relatively
free of arbitrary constraints. Think of how you process information. If
I give you the first line of a poem and I ask you who wrote the poem,
you don't keep a linear table in your head of all the first lines of all the
poems you ever heard with a pointer to an alphabetical table of authors.
Yet that is the way we think about our problems because we know
FORTRAN or Pascal.

18

THE WESTINGHOUSE NETWORK

PETER ZAPHYR

The Westinghouse Electric Corporation was founded in 1886 by George Westinghouse in Pittsburgh. In 1986 we celebrated our 100th anniversary as a corporation. The founder and the corporation were pioneers in electric power—its generation, distribution, and application. The corporation also pioneered numerous areas related to electric power such as commercial broadcasting, applications of advanced radar technology, and nuclear power. For example, Westinghouse produced the nuclear engine for the first nuclear submarine, the *Nautilus*, and the first commercial application of nuclear power for generation of electricity at the Shippingport, Pennsylvania, plant.

Today, Westinghouse is a $10 billion corporation; it is the 30th largest in sales of Fortune 500 U.S. industrial corporations, employing 127,000 people worldwide. The major businesses of Westinghouse include products and services in the energy, defense, industry, broadcasting, and credit markets. In addition, Westinghouse provides a broad range of commercial products and services such as land development, furniture systems, and bottling of soft drinks.

COMPUTERS IN WESTINGHOUSE

Since Westinghouse concentrated much of its efforts in high technology products from its earliest days, it was natural that Westinghouse would be among the first to employ computers in engineering. From the early 1930s, analog computers were constructed by Westinghouse engineers for such special applications as electric utility systems studies. In the late 1940s Westinghouse designed and built a general-purpose analog computer called the ANACOM that is still in active use today for calculating switching surges on electric power systems.

Westinghouse began employing digital computers as soon as they became commercially available in the early 1950s. From those beginnings Westinghouse proceeded to acquire and employ the most advanced digital computers commercially available throughout the following 30 years. These included the IBM series starting with the 701, the UNIVAC I and its successors, the PHILCO Transac S2000, the Burroughs B5000 series, the Control Data Corporation 6600 and the 7600, and more recently the Cray–1S and the CDC/ETA 205.

At present, we find in Westinghouse that analog computers are still useful, but limited to applications primarily in design, analysis, and simulation of control systems. Westinghouse has several major digital computer centers. It also uses clusters of diverse minicomputers throughout all of the business units and rapidly growing numbers of work stations. These computers and work stations are linked together through a data communications network connecting virtually every location of the corporation. Through that network the typical engineer has at his or her fingertips access to immense resources of computational power. We are currently working to improve that access by employing high-speed bus technology for more capable computer-to-computer networking. Thus supercomputers are but specialized components of a complex computer utility extending worldwide throughout the corporation.

THE WESTINGHOUSE SOFTWARE PORTFOLIO

Early engineering applications of digital computers in Westinghouse cover virtually every field of engineering. Some of the more prominent applications, by category, include:

- Mechanical: static and dynamic performance of generator rotors, steam-turbine shafts and blades, jet engines, and pumps
- Electrical and electronics: power-transformer design, electric utility systems analysis, motor design, switchgear design, control system analysis, and radar design
- Hydraulics: turbine-blade design, steam-turbine heat balance calculations, and nuclear-reactor fluid-flow analysis
- Nuclear power: diffusion calculations for power distribution and fuel depletion
- Manufacturing: numerical control and tool design

Computers changed the way Westinghouse did business, as illustrated by the 1954 development of a program for the design of power transformers. That program was intended to replace the engineer on customer-order design work, a concept that was heretical at the time. The work was successful and it is still employed today in Westinghouse. The transformer design program was recognized at the time by the American Institute of Electrical Engineers as pioneering work in the application of computers to engineering.

Another example was the development of a generalized computer program for calculating the performance of steam turbine systems used for the production of electric power. The technique is called the heat-balance calculation. That work had important engineering benefits, and also was employed commercially to size steam turbines during negotiations with electric utilities. The program continues today as a cornerstone in the applications portfolio of Westinghouse.

The reason for citing these early applications of digital computers to engineering is that virtually all of them achieved a life far beyond the expectations of those who originally built the programs. Through such efforts, thousands of engineering computer programs exist today in Westinghouse, forming a major asset of the corporation.

The current application software portfolio can best be characterized as having evolved over the past three decades by building on the original base of applications. That was done by refining design applications based on field experience in the operation of our products, by introducing more generalization in the structure of the programs, and by providing the more refined representations advanced computers permit. Also, through more powerful computers, the ability to optimize the design of products became possible. Substantial enrichments of the applications were introduced by employing interactive graphics and interactive computing techniques.

As computer programs captured the essence of the design process, it became necessary to impose a system for ensuring that changes to the design programs were introduced in a quality manner. Also, configuration management was applied to assure that those changes were maintained in a traceable manner in the event a design had to be recreated. Security measures were instituted to assure that the programs were protected from unauthorized access and protected against loss. Data-management techniques also became valuable as historical engineering data were developed and introduced into a computational library of information critical to the design and maintenance of our products.

Another significant development was the recognition that this growing application portfolio had substantial commercial value. Several divisions within Westinghouse market these applications, generally as value-added services. For example, we have today structural analysis for nuclear piping systems carried out in mobil units containing minicomputers for the interactive graphics work to capture as-built data. The minis are linked back to the central computers, such as the Crays, for the analysis work. Opportunities in application of artificial-intelligence techniques are being pursued—for example, in monitoring turbine generators during operation.

These commercial opportunities for computer-based services are reorienting our operational priorities in the computer center. A stronger emphasis is placed on providing stable services, high availability, and good responsiveness. Also, we have need for more management visibility of our revenue-producing computer services.

SUPERCOMPUTERS

Two installations within Westinghouse today have supercomputers. The Energy Systems Computer Center (ESCC) has two Cray–1S computers. These two supercomputers are integrated into a worldwide network and can be accessed via a large range of distributed smaller computers. Their function is to provide the engineering computer services needed to support Energy Systems and Defense organizations within Westinghouse. The second large computer center having supercomputers is at the Bettis Atomic Power Laboratories, which currently have a CDC/ETA 205 and four CDC 7600s. Their function is to provide technical services to the Navy nuclear propulsion program managed by Westinghouse. Figures 18.1 and 18.2 show the ESCC configuration and the data communications networking making those computers available to all parts of the corporation.

Westinghouse expects to have an ongoing need for the most capable general-purpose supercomputer available at any point in time. That will be necessary for us to continue to improve the accuracy of representations of our highly technical products and systems toward gaining commercial leverage through improved product quality and performance. For the computer center in the Energy and Advanced Technology group, we will probably require two such compatible supercomputers at any point in time in order to provide adequate backup and availability.

The lower end of the application base will be shifting toward the smaller interactive computers and work stations that are distributed on the network. This may lead to more specialization of applications on the supercomputers in the future.

I also expect the supercomputers that we operate to remain in a batch-processing mode, that is, in a non-interactive mode of operation. As the smaller applications move off to the interactive minis and work stations on the network, and as the larger applications continue to grow in demand for computer resources, I see little benefit to be obtained by introducing interactive capabilities for the long-running applications.

Other expectations are that we would continue to make more use of bus technology for networking the supercomputers into an integrated

Figure 18.1
ESCC Engineering Computer Configuration

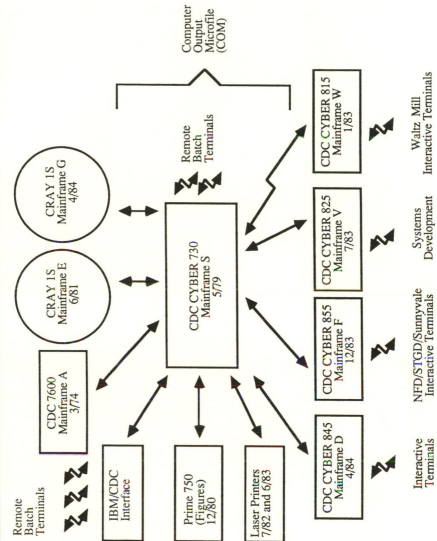

Figure 18.2
ESCC Terminal Network

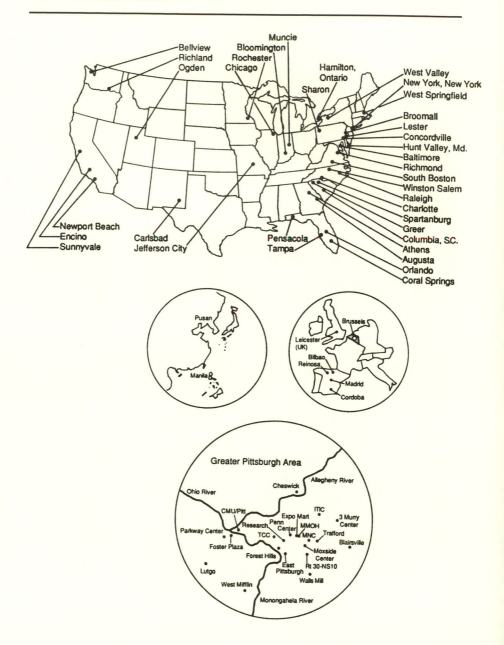

Source: Author.

corporate computer utility. Finally, I expect the trend of supercomputers as a means for producing revenues for Westinghouse to accelerate.

INHIBITORS TO PROGRESS

Most of these trends mentioned above have supported an acceleration in the use of supercomputers within Westinghouse. However, there are certain inhibitors to that continued progress. First is the great difficulty of converting from one supercomputer to another, particularly since there has been little emphasis in the past on maintaining compatibility from vendor to vendor and even within a single vendor's supercomputer product lines. Conversion of the large number of computer applications in our portfolio simply takes much too long. It takes too much effort and consumes too many valuable resources to allow us to convert from machine to machine very frequently.

We have gone through this process about every five years. As the application base continues to grow, and as the demands for application stability continue to increase, the burden of moving "the mountain to Muhammad" every five years is becoming unthinkable as a continuing way of operation.

Further, some of the changes introduced by computer hardware and operating systems designers appear to be without solid foundation from the user standpoint—for example, the introduction of nonstandard features and extensions to compilers such as FORTRAN. Another important inhibitor to continued progress in the effective use of supercomputers is the fact that there are limited numbers of vendors who are participating in the supercomputer market, while the prominent companies in the computer field have conspicuously avoided that market. If the supercomputer market will in the future be satisfied by vendors who specialize only in supercomputers, I am concerned that the practices that currently inhibit the growth of supercomputers will continue.

WESTINGHOUSE-UNIVERSITY COLLABORATION

Westinghouse has cooperative programs with the University of Pittsburgh and Carnegie-Mellon University. The programs have provided some small amount of Cray–1 computer time to the University of Pittsburgh for use in a certificate program at the master's level on the use of supercomputers in science. We also have been working with investigators and researchers at Carnegie-Mellon and Pitt to foster better understanding of the supercomputer field. This has taken the form of a number of jointly organized seminars. Westinghouse also provided technical support and expertise to joint efforts between Carnegie-Mellon University and Pitt in pursuing the establishment of a supercomputer center for the National Science Foundation located in the Pittsburgh area. That supercomputer center is now functioning and is being operated by Westinghouse for the universities.

SUMMARY

Westinghouse will continue to be an important user of supercomputers. At any point in time we expect not more than four or five supercomputers to be installed within the corporation. The ability to employ supercomputers productively depends upon the architects and designers of supercomputer hardware and software and their progress toward providing a stable interface between their designs and the applications portfolios of their users. Finally, both industry and government must work with the universities in promoting research and understanding in the use of supercomputers throughout their faculties and in their student bodies.

BIBLIOGRAPHY OF LITERATURE ON SUPERCOMPUTERS

JAMES C. BROWNE, JOHN FEO, and PATRICIA ROE

This Bibliography is an attempt to compile a focused summary of the current literature on supercomputers and their applications. Selection of articles to include was subjective; the main criterion for selection was relevance to the concept of the most powerful computers that can be built and applied in a given problem environment. It is divided into four categories: Applications, Architectures, Business and Marketing Aspects, and General Articles and Surveys.

We are grateful to Cray Research for providing us with their literature survey. Support for preparation of this Bibliography has been provided by the RGK Foundation of Austin, Texas, the IC² Institute, The University of Texas at Austin, and the Westinghouse Electric Corporation.

APPLICATIONS

Adelantado, M., D. Comte, P. Siron, and J. C. Syre. "An MIMD Supercomputer System for Large-Scale Numerical Applications." In *Proceedings of the IFIP 9th World Computer Congress*, 821–826. Amsterdam, Netherlands: North Holland, 1983.

Anderson, W., R. Brice, and W. Alexander. *Configuring a Supercomputer for an Interactive Scientific Workload*. Technical Report LA-UR–82–1572, Los Alamos National Laboratory, 1982.

Ballhaus, W. F., Jr. "Supercomputing in Aerodynamics." In *Frontiers of Supercomputing*. Berkeley: University of California Press, 1986.

Ballhaus, W. R. *Computational Aerodynamics and Supercomputers*. Technical Report NASA-TM–85887, NASA, 1984.

Bardon, M., and K. K. Curtis. *A National Computing Environment for Academic Research*. Technical Report, National Science Foundation, 1983.

Barlow, R. H., D. J. Evans, J. Shanenchi. "Comparative Study of the Exploitation of Different Levels of Parallelism on Different Parallel Architectures." In *Proceedings of the 1982 International Conference on Parallel Processing*, 34–40. Silver Spring, MD.: IEEE Computer Society Press, August, 1982.

Bersohn, M. "Supercomputers and the Problem of Organic Synthesis." In *ACS Symposium Series*, 109–116. Washington, D.C.: American Chemical Society, 1981.

Bishop, M., D. Ceperley, H. L. Frisch, and M. H. Kalos. "The Need for Supercomputers in Time-Dependent Polymer Simulations." In *ACS Symposium Series*, 135–142. Washington, D.C.: American Chemical Society, 1981.

Boris, J. P., and N. K. Winsor, "Vectorized Computation of Reactive Flow." In *Parallel Computations, Vol. 1, Computational Techniques*, 172–214. New York: Academic Press, 1982.

Bowler, K. C., and G. S. Pauly. "Molecular Dynamics and Monte Carlo Simulations in Solid State in Elementary Particle Physics." In *Proceedings of the IEEE 72*, 42–55. IEEE, 1984.

Buzbee, B. L. "Gaining Insight from Supercomputing." In *Proceedings of the IEEE 72*, 19–21. IEEE, 1984.

Byrne, G. W. "A View of Supercomputing." In *Frontiers of Supercomputing*. Berkeley: University of California Press, 1986.

Calahan, D. A. "A Block-Oriented Sparse Equation Solver for the Cray-1." In *Proceedings of the 1979 International Conference on Parallel Processing*, 116–123. New York: IEEE Computer Society Press, 1979.

Collins, J. R., and A. F. Caulson. "Applications of Parallel Processing Algorithms for DNA Sequence Analysis." *Nucleic Acids Research* 12 (1984): 181–92.

Cromie, W. J., and L. Edson. "Before They Can Speak, They Must Know." *Mosaic* 15 (1984): 28–35.

Davis, C. G., and R. L. Couch. "Ballistic Missile Defense: A Supercomputer Challenge." *Computer* 13 (1980): 37–46.

Demos, G., M. D. Brown, and R. A. Weinberg. "Digital Scene Simulation: The Synergy of Computer Technology and Human Creativity." In *Proceedings of the IEEE 72*, 22–31. IEEE, 1984.

Dennis, J. B., G. R. Gao, and K. W. Todd. "Modeling the Weather with a Data Flow Supercomputer." *IEEE Transactions on Computers C–33* (1984): 592–603.

Department of Energy. *The Role of Supercomputers in Energy Research Programs*. Washington, D.C.: Department of Energy, 1985.

Doyle, J. *What Should A.I. Want from the Supercomputers?* Technical Report CS–83–160, Carnegie-Mellon University, 1983.

Dubois, P. F. "Swimming Upstream: Calculating Table Lookups and Piecewise Functions." In *Parallel Computations, Vol. 1, Computational Techniques*, 129–151. New York: Academic Press, 1982.

Egan, J. T., and R. D. MacElroy. "Calculating the Electrostatic Potential of Molecular Models with Separate Evaluations by Conventional, Vector and Array Processors." *Journal of Computational Chemistry* 5 (1984): 52–81.

Elson, B. M. "Supercomputer Sought to Aid Design." *Aviation Week*, April 14, 1978, pp. 123–126.

Evans, D. J., H. J. M. Hanley, and S. Hess. "Non-Newtonian Phenomena in Simple Fluids." *Physics Today* 37 (1984): 26–35.

Fichtner, W., L. W. Nagel, B. R. Penumalli, W. P. Peterson, and J. L. D'arcy.

"The Impact of Supercomputers on IC Technology Development and Design. In *Proceedings of the IEEE* 72, 96–112. IEEE, 1984.

Fox, G. C., and S. W. Otto. "Algorithms for Concurrent Processors." *Physics Today* 37 (1984): 53–65.

Fuss, D., and C. G. Tull. "Centralized Supercomputer Support for Magnetic Fusion Energy Research." In *Proceedings of the IEEE* 72, 32–41. IEEE, 1984.

Garfinkel, D. "We Could Wire Up an Intelligent Artificial Pancreas—But What Would We Tell It to Do?" *Perspectives in Computing* 43 (1984): 10–16.

Gloudeman, J. F. "Anticipated Impact of Supercomputers and Finite Element Analysis." In *Proceedings of the IEEE* 72, 80–89. IEEE, 1984.

Graedel, T. E., and R. McGill. "Graphical Presentation of Results from Scientific Computer Models." *Science*, March 1982, 1191–1198.

Guest, M. F., and S. Wilson. "The Use of Vector Processors in Quantum Chemistry." *ACS Symposium Series*, 1–37. Washington, D.C.: American Chemical Society. 1981.

Happ, H. H. "Parallel Processing in Power Systems." In *Proceedings of the 7th Power System Computing Conference*, 9–16. Guildford, England: Westbury House, July 1981.

—————. "Parallel Processing in Power Systems." *International Journal of Electrical Power and Energy Systems* 4 (1982): 37–41.

Harlow, F. H. "Superproblems for Supercomputers." In *Frontiers of Supercomputing*. Berkeley: University of California Press, 1986.

Heine, R. W. "CAD/CAM: Ford Motor Company Engineering and Manufacturing." In *Frontiers of Supercomputing*. Berkeley: University of California Press, 1986.

Henderson, D. "Computation: The Nexus of Nuclear Weapon Development." In *Frontiers of Supercomputing*. Berkeley: University of California Press, 1986.

Herr, R. A. "Computer Models Gaining on El Nino." *Science*, July 1984, 37–38.

Hood, R. T., and K. Kennedy. "Programming Language Support for Supercomputers." In *Frontiers of Supercomputing*. Berkeley: University of California Press, 1986.

Hoover, W. G. "Computer Simulation of Many-Bodied Dynamics." *Physics Today* 37 (1984): 44–50.

Horton, E. J., and W. D. Compton. "Technological Trends in Automobiles." *Science*, August 1984, 587–930.

Huff, R. W., J. M. Dawson, and G. J. Culler. "Computer Modeling in Plasma Physics on the Parallel-Architecture CHI Computer." In *Parallel Computations, Vol 1, Computational Techniques*, 365–396. New York: Academic Press, 1982.

Ishiguro, M., H. Harada, et al. *Applicability of Vector Processing to Large-Scale Nuclear Code*. Technical Report JAERI-M–82–018, Japan Atomic Energy Research Institute, 1982.

James, R. A. "Simulation of Particle Problems in Astrophysics." *Computer Physics Communications* 26 (1982): 412–431.

Johnson, O. G. "Three-Dimensional Wave Equation Computations on Vector Computers." In *Proceedings of the IEEE* 71, 90–95. IEEE, 1984.

Karplus, W. J. "The Impact of New Computer System Architectures on the

Simulation of Environmental Systems." In *Conference on Modeling, Identification, and Control in Environmental Systems*, 1001–1009. Amsterdam: North Holland, 1977.

Kershaw, D. "Solution of Single Tridiagonal Linear Systems and Vectorization of the ICCG Algorithm on the Cray–1." In *Parallel Computations, Vol. 1, Computational Techniques*, 85–90. New York: Academic Press, 1982.

Killough, J. E. "The Use of Vector Processors in Reservoir Simulation." In *Proceedings 5th Symposium on Reservoir Simulation*, 21–28, Dallas: Society Petroleum Engineers of AIME, 1979.

Kincaid, D. R., and T. C. Oppe. "ITPAK on Supercomputers." In *Proceedings of the 1982 International Workshop on Numerical Methods*, 151–61. Berlin-New York: Springer-Verlag, June 1982.

Kolata, G. "A Fast Way to Solve Hard Problems." *Science*, September 1984, 1379–1380.

Krzeczlowski, A. J., E. A. Smith, and T. Gethin. "Seismic Migration Using the ICL Distributed Array Processor." *Computer Physics Communications* 26 (1982): 447–53.

Kuck, D. J. "ILLIAC IV Software and Application Programming." *IEEE Transactions on Computers* C–17 (1968): 758–70.

Lomax, H., and T. H. Pulliam. "A Fully Implicit, Factored Code for Computing Three-Dimensional Flows on the ILLIAC IV." In *Parallel Computations, Vol. 1, Computational Techniques*, 217–249. New York: Academic Press, 1982.

Lykos, P., and I. Shavitt. "Supercomputers in Chemistry." *ACS Symposium Series*. Washington, D.C.: American Chemical Society, 1981.

MacCormack, R. W., and K. G. Stevens. "Fluid Dynamics Applications of the ILLIAC-IV Computer." In *Proceedings of the Computation Methods and Problems in Aeronautical Fluid Dynamics*, 448–465. London: Academic Press, 1974.

Maples, C., D. Weaver, W. Rathbun, and J. Meng. "The Utilization of Parallel Processors in a Data Analysis Environment." *IEEE Transactions on Nuclear Science* NS–28 (1981): 3880–3888.

"Massively Parallel Processor Yields High Speed." *Aviation Week & Space Technology*, May 18, 1984, 157.

McCoy, M. G., A. A. Mirin, and J. Killeen. *Vectorized Fokker-Planck Package for the Cray–1*. Technical Report CON–790902–1, Lawrence Livermore National Laboratory, 1979.

McKerrell, A., and L. M. Delves. "Solution of the Global Element Equations on the ICL DAP." *ICL Technical Journal* 4 (1984): 50–58.

McLean, W. J. *Report of the Ad-Hoc Combustion Research Facility Commission on Computational Resources for Combustion Research*. Technical Report SAND–83–8237, Sandia National Laboratory, 1983.

"National Security Aspects of Supercomputer R&D." *Computer Age*, December 19, 1983, 1, 4.

Nelson, D. B. "Supercomputers and Magnetic Fusion Energy." *Frontiers of Supercomputing*. Berkeley: University of California Press, 1986.

Nelson, H. R. "New Vector Supercomputers Promote Seismic Advancements." *World Oil*, January 1982, 155–60.

Newton, A. R., and D. O. Pederson. *Research in Computer Simulation of Integrated Circuits.* Technical Report AFOSR–82–0021, Air Force Office of Scientific Research, 1983.

Norrie, C. "Supercomputers for Superproblems: An Architectural Introduction." *Computer*, March 1984, 62–74.

Orme, M. "The Ultimate in CAD Tools." *Technology* 8 (1984): 24–25.

Ostlund, N. S., P. G. Hibbard, and R. A. Whiteside. "A Case Study in the Application of a Tightly Coupled Multiprocessor to Scientific Computations." In *Parallel Computations, Vol. 1, Computational Techniques*, 315–63. New York: Academic Press, 1982.

Parkinson, D. "Using the ICL DAP." *Computer Physics Communications* 26 (1982): 227–232.

Parkinson, D. and H. M. Liddell. "The Measurement of Performance on a Highly Parallel System." *IEEE Transactions on Computers* C–32 (1983): 32–37.

Parkinson, D., and J. Sylwestrowicz. "DAP in Action." *ICL Technical Journal* 3 (1983): 330–344.

Peterson, V. L. *Application of Supercomputers to Computational Aerodynamics.* Technical Report NASA-TM–85965, NASA, 1984.

————. "Impact of Computers on Aerodynamics Research and Development." In *Proceedings of the IEEE* 72, 68–79. IEEE, 1984.

Potter, J. L. "Image Processing on the Massively Parallel Processor." *Computer* 16 (1983): 62–67.

Pottle, C. "Solution of Sparse Linear Equations Arising from Power System Simulation on Vector and Parallel Processors." *ISA Transactions* 18 (1979): 81–88.

"RNA Modeling for Biotechnology." *Cray Channels*, Spring 1985, 8–11.

Rodrigue, G., C. Hendrickson, and M. Pratt. "An Implicit Numerical Solution of the Two-Dimensional Diffusion Equation and Vectorization Experiments." In *Parallel Computations, Vol. 1, Computational Techniques*, 101–128. New York: Academic Press, 1982.

Rose, D. J. "Numerical Analysis for VLSI Simulation: Pointers and Reflections," *Frontiers of Supercomputing.* Berkeley: University of California Press, 1986.

Schatz, W., and J. W. Verity. "DARPA's Big Push in A.I." *Datamation* 30 (1984): 48–50.

Schommer, N. "Crashing Images." *Discover*, October 1984, 74–76.

Schwartz, J. T. "Mathematical Problems in Robotics." In *Frontiers of Supercomputing.* Berkeley: University of California Press, 1986.

Siebel, G. "Computational Chemistry: A Biomacro Molecule." *Cray Channels*, Spring 1985, 3–7.

Solem, J. C. "*MECA: A Supercomputer for Monte Carlo.*" Technical Report LA–10005, Los Alamos National Laboratory, 1984.

Stevenson, D. K. "Numerical Algorithms for Parallel Computers." In *1980 AFIPO Conference Proceedings*, 357–361, May 1980.

Strikwerda, J. C. "A Time-Split Difference Scheme for the Compressible Navier-Stokes Equations with Applications to Flows in Slotted Nozzles." In *Parallel Computations, Vol. 1, Computational Techniques*, 251–67. New York: Academic Press, 1982.

Swarztrauber, P. N. "Vectorizing the FFTs." In *Parallel Computations, Vol. 1, Computation Techniques*, 51–84. New York: Academic Press, 1982.

"Systems & Peripherals: MPP Processes Satelite Data." *Computerworld*, February 13, 1984, 99.

Teitelman, R., "Designs for Living." *Forbes*, November 1984, 298–302.

Thapar, N. "Solving a Drug Problem. *Computering*, December 1984, pp. 19–20.

Walker, R. B., P. J. Hay, and H. W. Galbraith. *Supercomputer Requirements for Theoretical Chemistry*. Technical Report CONF–800814–26, Los Alamos National Laboratory, 1980.

————. "Supercomputer Requirements for Theoretical Chemistry." In *ACS Symposium Series*, 47–63. Washington, D.C.: American Chemical Society, 1981.

Welsh, J. G. "Geophysical Fluid Simulation on a Parallel Computer." *Parallel Computations, Vol. 1, Computational Techniques*. 169–177. New York: Academic Press, 1982.

Willerton, D. L. *Cray/VAX Interactive Graphics System*. Technical Report LA–UR– 83–3397, Los Alamos National Laboratory, 1983.

Williamson, D. L., and P. Swarztraber. "A Numerical Weather Prediction Model-Computational Aspects on the Cray–1." In *Proceedings of the IEEE 72*, 56–57. IEEE, 1984.

Wilson, K. G. "The Role of Universities in Very High Performance Computing." *Frontiers of Supercomputing*. Berkeley: University of California Press, 1986.

————. "Experiences with a Floating Point Systems Array Processor." In *Parallel Computations, Vol. 1, Computational Techniques*, 279–313. New York: Academic Press, 1982.

Woodward, P. R. "Trade-Offs in Designing Explicit Hydrodynamical Schemes for Vector Computers." In *Parallel Computations, Vol. 1, Computational Techniques*, 153–171. New York: Academic Press, 1982.

Wood, L. L. *Conceptual Basis for Defense Applications of Supercomputers*. Technical Report UCID–19746, Lawrence Livermore National Laboratory, 1983.

ARCHITECTURES

Alexander, G. "SSI/MSI/LSI/VLSI/USLI." *Mosaic* 15 (1984): 10–17.

Alexander, T. "Computing with Light at Lightning Speed." *Fortune International*, July 23, 1984, 74–78.

"And Now, an 'Affordable Supercomputer.' " *Business Week Industrial Edition*, November 1984, 164.

Babb, R. G. "Parallel Processing with Large-Grain Data Flow Techniques." *Computer* 17 (1984): 55–61.

Baer, J. L. *Computer Systems Architecture*. Potomac, MD: Computer Science Press, 1980.

————. "Multiprocessing Systems." *IEEE Transactions of Computers* C–25 (1976): 1271–1277.

Baqai, I. A., and T. Lang. "Reliability Aspects of the ILLIAC-IV Computer." In *Proceedings of the 1976 International Conference on Parallel Processing*, 123–121. New York: IEEE Computer Society, 1976.

Barnes, G. H., R. M. Brown, et al. "The ILLIAC IV Computer." *IEEE Transactions on Computers* C17 (1968): 746–757.

Batcher, K. "MPP-A Massively Parallel Processor." In *Proceedings of the 1979 International Conference on Parallel Processing*, New York: IEEE Computer Society Press, 1979.

Broughton, J. M., P. M. Farmwald, and T. M. Mcwilliams. *S–1 Multiprocessor System*. Technical Report UCRL–87494, Lawrence Livermore National Laboratory, 1982.

Browne, J. C. "Parallel Architectures for Computer Systems." *Physics Today* 37 (1984): 28–37.

Bucher, I. Y., and J. W. Moore. *Comparative Performance Evaluation of Two Supercomputers: CDC Cyber–205 and CRI Cray–1*. Technical Report LA-UR–81–1977, Los Alamos National Laboratory, 1981.

Budnik, P., and D. J. Kuck. "The Organization and Use of Parallel Memories." *IEEE Transactions of Computers* C-20 (1971): 1566–1569.

Davis, A. L. "Computer Architecture." *Spectrum*, November 1983, 94–99.

Davis, R. L. "The ILLIAC IV Processing Element." *IEEE Transaction on Computers* C–18 (1969): 800–816.

Dennis, J. B. "Data Flow Ideas in Future Supercomputers." In *Frontiers of Supercomputing*. Berkeley: University of California Press, 1986.

————. "Data-Flow Supercomputers." *Computer* 13 (1980): 48–56.

Drogin, E. M. "Pipeline Architecture Battles Array." *MSN Microwave System News*, October 1980, 92–96.

Efe, K. *Task Allocation in Supersystems*. Technical Report LU/DCS/R–173, Leeds University, 1983.

Farmwald, P. M., W. Bryson, and J. L. Manferdelli. *Signal Processing Aspects of the S–1 Multiprocessor Project*. Technical Report CONF–800719–7, Lawrence Livermore National Laboratory, 1980.

Gottlieb, A. "The New York University Ultracomputer." *Frontiers of Supercomputing*. Berkeley: University of California Press, 1986.

Hanuliak, I. "Computer Systems from the Viewpoint of Architecture." *Automatizace* 23 (1980): 205–209.

Hiatt, B., and P. Gwinne. "Parceling the Power." *Mosaic* (1984): 36–43.

Hogen, D. J. "Computer Performance Prediction of a Data-Flow Architecture." Master's thesis, Naval Postgraduate School, 1981.

Hoshind, T., T. Kageyama, T. Shirakawa, et al. "Highly Parallel Processor Array 'PAX' for Wide Scientific Applications." In *Proceedings of the 1983 International Conference on Parallel Processing*, 95–105. Silver Spring, MD: IEEE Computer Society Press, 1983.

————. "PACS: A Parallel Microprocessor Array for Scientific Calculations." *ACM Transactions on Computer Systems* 1 (1983): 195–221.

Jensen, C. "Taking Another Approach to Supercomputing." *Datamation*, February 1978, 156–172.

Jordan, T. L. "A Guide to Parallel Computation and Some Cray–1 Experiences." In *Parallel Computations, Vol. 1, Computational Techniques*, 1–49. New York: Academic Press, 1982.

Kaneda, Y. "Parallel Processors." *Systems and Control* 27 (1983): 508–516.

Karplus, W. J. "Parallelism and Pipelining in High-Speed Digital Simulators."

In *10th IMACS World Congress on Systems Simulation Scientific Computation*. 272–274. New Brunswick, N.J.: IMACS, August 1982.

Kartashev, S. I., and S. P. Kartashev. "Problems of Designing Supersystems with Dynamic Architectures." *IEEE Transactions on Computers* C–29 (1980): 1114–1132.

Kawanobe, K. "Present Status of the Fifth Generation Computer Systems Project." *ICOT Journal* 5 (1984): 13–21.

Kozdrowicki, E. W. "Supercomputers for the Eighties." *Digital Design* 13 (1983): 94–103.

Kozdrowicki, E. W., and D. J. Theis. "Second Generation of Vector Supercomputers." *Computer* 13 (1980): 71–83.

Kuck, D. J., D. Laurie, R. Cytron, H. A. Same, and D. Gajski. "Cedar Project." *Frontiers of Supercomputing*. Berkeley: University of California Press, 1986.

————— and D. A. Padua, "High-Speed Multiprocessors and Their Compilers." In *Proceedings of the 1979 International Conference on Parallel Processing*, 5–16. New York: IEEE Computer Society Press, August 1979.

Kung, H. T. "Special Purpose Supercomputers." Technical Report, Carnegie-Mellon University, 1984.

Labrecque, M. "The Many Ways Data Must Flow." *Mosaic* 15 (1984): 18–27.

Levine, R. D. "Supercomputers." *Scientific American*, January 1982, 118–135.

Manuel, T. "Advanced Parallel Architectures Get Attention as Way to Faster Computing." *Electronics*, June 16, 1983, 105–106.

—————. "Cautiously Optimistic Tone Set for 5th Generation." *Electronics Week*, December 3, 1984, 57–63.

Martin, H. G. "A Discourse of a New Super Computer, PEPE." In *Conference on High Speed Computer and Algorith Organization*, 101–112. London: Academic Press, April 1977.

Parkinson, D. "The Distributed Array Processor (DAP)." *Computer Physics Communications 28* (1983): 325–336.

"A Plethora of Projects in the U.S. Try Data-flow and Other Architectures." *Electronics*, June 16, 1983, 107–110.

Requa, J. E., and J. R. McGraw. "The Piecewise Data Flow Architecture: Architectural Concepts." *IEEE Transactions on Computers* C–32 (1983): 425–38.

Rudy, T. E. "Megaflops from Multiprocessors." In *Proceedings of the 2nd Rocky Mountain Symposium on Microcomputers*, 99–107. New York: IEEE Computer Society Press, August 1978.

Sahni, S. *Scheduling Supercomputers*. Technical Report TR–83–3, University of Minnesota, 1983.

Schindler, M. "Multiprocessing Systems Embrace Both New and Conventional Architectures." *Electronic Design* 32 (1984): 97–130.

Seitz, C., and J. Matisoo. "Engineering Limits on Computer Performance." *Physics Today 37* (1984): 38–52.

Simmons, G. H. "TDMP: A Data Flow Processor." PhD thesis, Michigan State University, 1981.

Smith, B. J., and D. J. Fink. "Architecture and Applications of the HEP Multiprocessor Computer System." In *Proceedings of the 1982 Conference on Peripheral Array Processors*, 159–170. La Jolla, CA: SCS, October 1982.

Smith, M. G., W. A. Notz, and E. Schischa. "The Question of Systems Implementation with Large-Scale Integration." *IEEE Transactions on Computers* C–18 (1969): 690–94.

Snyder, L. *Supercomputers and VLSI: The Effect of Large Scale Integration on Computer Architecture.* Technical Report TR–834–08–05, University of Washington at Seattle, 1984.

Su, S. P. "Pipelining and Dataflow Techniques for Designing Supercomputers." Ph.D. dissertation, Purdue University, 1982.

"SuperCPUs: Japanese Close In." *Electronic Engineering Times*, March 16, 1984, 1, 12.

Swartzlander, E. E., and B. K. Gilbert. "Supersystems: Technology and Architecture." *IEEE Transactions on Computers* C–31 (1982): 399–409.

Thurber, K. J. "High-Performance Parallel Processors." In *Proceedings of the Society of Photo-Optical Instrument Engineers*, 45–59. Bellingham, WA: Society Photo-Optical Engineers, August 1978.

Treleaven, P. C. "Future Computers: Logic, Data Flow, Computer Flow?" *Computer* 17 (1984): 47–55.

Vick, C. R., Kartashev, S. P., and S. I. Kartashev. "Adaptable Architectures for Supercomputers." *Computer* 13 (1980): 17–35.

Widdoes, L. C. *S–1 Project: Developing High-Performance Digital Computers.* Technical Report CONF–800201–5, Lawrence Livermore National Laboratory, 1980.

BUSINESS AND MARKETING ASPECTS

Alexander, J. "Reinventing the Computer (Parallel Processing)." *Fortune*, March 5, 1984, 62–70.

Alexander, T. "Cray's Way of Staying Super-duper." *Fortune*, March 18, 1985, 66–68, 72, 76.

"Avionics: Smaller Firms Dominate Supercomputer Field in U.S." *Aviation Week & Space Technology*, May 28, 1984, 154–157.

Benoit, E. "Filling the Gap." *Forbes*, March 1985, 166–170.

Carlyle, R. E. "Here Come the Crayettes." *Datamation* 31 (1985): 40–42.

Cook, J. "War Games." *Forbes*, September 12, 1983, 108, 110.

Corrigan, R. "Inman's Innovation." *National Journal*, March 5, 1983, 523.

Feigenbaum, E. A., and P. McCorduck. "Japan's MITT Strike." *Savvy*, May, 1983, 38, 40, 42, 44.

Fernbach, S. *The IEEE Supercomputer Committee Report.* Technical Report IEEE, 1983.

"A Fifth Generation: Computers That Think." *Business Week*, December 14, 1981, 94–96.

"Industry's Use of Computers Shows Big Increase." *Oil & Gas Journal*, April 9, 1984, 41–44.

Johnson, J. "Cray and CDC Meet the Japanese." *Datamation* 30 (1984): 32–42.

Kartashev, S. P., and S. I. Kartashev. "Supersystems for the 80's." *Computer* 13 (1980): 11–14.

Kurita, S. "Supercomputers Battle Japan Inc. on Its Own Turf." *Electronics Business*, December 10, 1984, 70–73.

Lineback, J. R. "Japan Looms as a Key Market for U.S.-made Supercomputers." *Electronics Week*, September 24, 1984, 18–19.

Manuel, T., and C. L. Cohen. "Computers of All Categories Increase in Performance: Fifth-Generation Machines and Supercomputers Hottest." *Electronics Week*, September 24, 1984, 58–60.

Mendez, R., and S. Orszag. "The Japanese Supercomputer Challenge." *Datamation*, May 15, 1984, 113, 114, 116, 119.

"Supercomputers Are Breaking Out of a Once Tiny Market." *Business Week Industrial Edition*, November 1984, 164.

Tyler, M. "Amdahl's Super CPU Gamble." *Datamation*, November 1, 1984, 36, 38, 40, 42.

Uttal, Bro. "Here Comes Computer Inc." *Fortune*, October 4, 1982, 82–90.

Wilson, K. G. "Science, Industry and the New Japanese Challenge." *Proceedings of the IEEE 72*, 6–18. IEEE, 1984.

GENERAL ARTICLES AND SURVEYS

Buzbee, B. L., R. H. Ewald, and W. J. Worlton. "Japanese Supercomputer Technology." *Science*, December 1982, 1189–1193.

Cohen, C. L. "Japanese Fifth-Generation Program Clicking Along Right on Schedule." *Electronics Week*, July 23, 1984, 33–34.

"The Coming Generation of Supercomputers." *Science Digest*, March 1983, 74–75.

Davidson, H. L. *Some Predictions on the Performance of Future Supercomputers for Simulation and Control.* Technical Report UCRL–89969, Lawrence Livermore National Laboratory, 1983.

Davis, D. B. "Supercomputers: A Strategic Imperative?" *High Technology*, May, 1984, 44, 46, 52.

Dickson, D. "Britain Rises to Japan's Computer Challenge." *Science*, May 20, 1983, 799–800.

Hendrickson, C. P. *Thinking Big.* Technical Report UCRL–90392, Lawrence Livermore National Laboratory, 1984.

Iversen, W. R. "Supercomputers Find New Jobs." *Electronics*, July 28, 1982, 57–76.

Kahn, R. E. "A New Generation in Computing." *Spectrum*, November 1983, 36–41.

Knight, J. C. "Current Status of Supercomputers." *Computers & Structures*, January 1979, 401–409.

Lemmons, P. "Japan and the Fifth Generation." *Byte*, November 1983, 394–401.

Malik, R. "Japan's Fifth Generation Computer Project." *Futures*, June 1983, 205–210.

Manuel, T. "Hollywood Signs on Supercomputer." *Electronics*, August 11, 1982, 56.

Marcom, J., and E. S. Browning. "Japan's Supercomputers: Breakthrough or Boast?" *Wall Street Journal*, August 8, 1984, 340.

"The Market for Advanced-Technology Computers." *Computer Age*, July 9, 1984, 1, 5.

Marsh, P. "The Race for the Thinking Machine." *New Scientist*, July 8, 1982, 85–87.

"NAS Enters Vector Processing Arena with Supercomputer." *Computerworld*, July 9, 1984, 8.

Norman, C. "House Votes Florida State a Supercomputer." *Science*, June 8, 1984, 1075–1076.

Oberdorfer, D. "Japan's New Challenges." *Washington Post*, July 23, 1984, 22.

Pehrs, J. "Rank and File Leap in Performance (Supercomputers Are Coming)." *Chip*, December 1983, 16–18.

"The Race is Still Open." *The Economist*, November 17, 1984, 88, 93.

Sanger, D. E. "The Surge in Supercomputers." *New York Times*, March 1, 1985, D1, D16.

Schefter, J. "Fifth-Generation Computers." *Popular Sciences*, April 1983, 79–82.

Seligman, M. "The Fifth Generation." *PC World*, October, 1983, 282, 284, 286.

Shimoda, H. "Fifth Generation Computer: From Dream to Reality." *Electronic Business*, November 1, 1984, 17.

Slotnick, D. L. "Centrally-Controlled Parallel Processors." In *Proceedings of the 1981 International Conference on Parallel Processing*. Silver Spring, MD: IEEE Computer Society Press, August 1981.

Smith, K. "U.K. Pursues Fifth-Generation Computer." *Electronics*, May 31, 1983, 101–102.

Solomon, S. "Superbrain: The Race to Create the World's Fastest Computer." *Science Digest*, September 1983, 42–49.

"Supercomputer Capacity in the Netherlands, An Inventory of Necessity and Availability." *Informatie*, April 1984, 294–301.

"Supercomputers Come Out into the World." *The Economist*, August 11, 1984, 77–80.

"Super Problems for Supercomputers." *Science News*, September 29, 1984, 200–203.

Torrero, E. A. "Tomorrow's Computers-The Quest" (special issue). *IEEE Spectrum*, November 1983.

Wallich, P. "Designing the Next Generation." *Spectrum*, November 1983, 73–77.

Walter, G. "Intelligent Supercomputers: The Japanese Computer Sputnik." *Journal of Information Image Management* 16 (1983): 18–22.

Walton, P., and P. Tate. "Soviets Aim for 5th Generation." *Datamation*, July 1, 1984, 53, 56, 61, 64.

"Western Europe Looks to Parallel Processing for Future Computers." *Electronics*, June 16, 1983, 111–113.

Wilson, K. G. "Supercomputer Powerful Tool." *Austin American Stateman*, April 21, 1985, J4.

Worlton, J. "Understanding Supercomputer Benchmarks." *Datamation*, September 1, 1984, 120, 122, 124, 128, 130.

Yasaki, E. K. "Japan Goes for the Gusto." *Datamation*, August 1981, 40, 44, 47, 50, 55.

INDEX

CONTRIBUTORS

F. Brett Berlin
Vice President
Government Relations
Cray Research

Dr. James C. Browne
David C. Bruton Centennial
   Professor of Computer Sciences
The University of Texas at Austin

Dr. James Decker
Deputy Director
Office of Energy Research
Department of Energy

Dr. George D. Dodd
Head
Computer Science Department
General Motors Research
   Laboratories

John Feo
Research Assistant
Computer Science Department
The University of Texas at Austin

Dr. Randolph Graves
Deputy Director
Aerodynamics Division
National Aeronautics and Space
   Administration

Dr. W. Daniel Hillis
Thinking Machines, Inc.

Dr. Glenn Ingram
Associate Director for Computing
National Bureau of Standards

Dr. Robert Johnson
Vice President, Graduate Studies and
    Research
Florida State University

Dr. Sidney Karin, Director
San Diego Supercomputer Center
GA Technologies
University of California, San Diego

Dr. John Killeen
National Magnetic Fusion Energy
    Computer Center
Lawrence Livermore National
    Laboratory

Dr. J. R. Kirkland
Principal
FBA, Inc.
Washington, D.C.

Dr. George Kozmetsky
Director
IC^2 Institute
The University of Texas at Austin

Dr. Joseph Lannutti
Director
Supercomputer Computations
    Research Institute
Florida State University

Dr. R. David Lowry
Director
Market Research and Development
Denelcor

Wayne McIntyre
Director
Special Purpose Systems
Amdahl Corporation

Dr. Walter McRae
Advanced Computational Center
University of Georgia

Edwin B. Neitzel
Group General Manager
Geo-Physical Research, ARCO

Dr. David Pensak
Supervisor
Computational Chemistry
Central Research Department
E.I. Dupont de Nemours Co.

Dr. J. H. Poore
Professor and Head, Department of
    Computer Science
University of Tennessee at Knoxville

Patricia Roe
Research Associate
IC² Institute
The University of Texas at Austin

Dr. Larry Smarr
Director
National Center for Supercomputer
    Applications
University of Illinois

Lloyd M. Thorndyke
President
Marketing Support
ETA Systems

Dr. Ken Wilson
James A. Weeks Professor of Physical
    Science
Cornell University

Henry Zanardelli
Manager
Engineering Computer Center
Ford Motor Company

Dr. Peter Zaphyr
Director
Energy Systems Computer Center
Westinghouse Electric Corporation

ABOUT THE EDITORS

J. R. KIRKLAND

J. R. Kirkland is a senior partner in FBA, Inc., a Washington-based government-relations firm that represents businesses, associations, and universities in Congress, at the White House, and in executive agencies.

Dr. Kirkland has a special interest and substantial experience in small growth companies and the public policies that affect this sector of the American business community. He was a member of the White House Domestic Policy Staff during the Carter administration and directed the White House Conference on Small Business. He has served as a consultant to the American Management Association and the Small Business Legislative Council, and is currently a member of the Small Business Council of the U.S. Chamber of Commerce.

In addition to his business activities, Dr. Kirkland has been active in both politics and education. He has served as Director of Research for the Democratic National Committee and was Executive Director of the Democratic Party of Georgia under Jimmy Carter. Dr. Kirkland, who holds a doctorate in history from the University of North Carolina, has been a professor at Cornell University, American University, and Heidelberg University in Germany. Currently he is Senior Research Fellow at the IC^2 Institute at The University of Texas at Austin and a visiting professor at Florida State University.

J. H. POORE, JR.

Jesse H. Poore, Jr. is Professor and Head of the Department of Computer Science at the University of Tennessee at Knoxville. Dr. Poore was a Professor of Information and Computer Science at the Georgia Institute of Technology, where he also held the administrative positions of Assistant to the President for Information Technology and Associate Vice President for Academic Affairs. In these capacities he planned, organized, and managed the communications and computing affairs of Georgia Tech. Under his leadership, Georgia Tech created an advanced computing environment featuring one of the largest university data-communications networks in the country.

Dr. Poore served as the Executive Director of the Committee on Science and Technology of the U.S. House of Representatives in the 98th Congress. In this capacity he became involved in general science policy activities and the supercomputer initiatives in particular. During the Carter administration he worked with the Office of Management and Budget under the executive loan program. In 1974 he was a Program Manager at the National Science Foundation. In addition to these positions, Dr. Poore has performed numerous voluntary services for agencies of government and is currently completing his second term as a member of the Panel for Scientific Computing of the NBS for the National Academy of Science/National Research Council.

Dr. Poore holds a doctorate in Information and Computer Science from the Georgia Institute of Technology and bachelor's and master's degrees in Mathematics. Professor Poore is a Senior Research Fellow of the IC^2 Institute of The University of Texas at Austin.

ABOUT THE SPONSORS

RGK FOUNDATION

The RGK Foundation was established in 1966 to provide support for medical and educational research. Major emphasis has been placed on the research of connective tissue diseases, particularly scleroderma. The Foundation also supports workshops and conferences at educational institutions through which the role of business in American society is examined. Such conferences have been co-sponsored with leading research and academic institutions.

The RGK Foundation Building has a research library and provides research space for scholars in residence. The building's extensive conference facilities have been used to conduct national and international conferences. Conferences at the RGK Foundation are designed not only to enhance information exchange on particular topics, but also to maintain an interlinkage among business, academia, community, and government.

IC² INSTITUTE

The IC² Institute at The University of Texas at Austin is a major research center for the study of Innovation, Creativity, and Capital (hence IC²). The Institute studies and analyzes information about the enterprise system through an integrated program of research, conferences, and publications.

IC² studies provide frameworks for dealing with current and critical unstructured problems from a private-sector point of view. The key areas of research and study concentration of IC² include: the management of technology; creative and innovative management; measuring the state of society; dynamic business development and entrepreneurship; econometrics, economic analysis, and management sciences; the evaluation of attitudes, opinions, and concerns on key issues.

The Institute generates a strong interaction between scholarly developments and real-world issues by conducting national and international conferences, developing initiatives for private and public-sector consideration, assisting in the establishment of professional organizations and other research institutes and centers, and maintaining collaborative efforts with universities, communities, states, and government agencies.

IC^2 research is published through monographs, policy papers, technical working papers, research articles, and three major series of books.